Phytates in Cereals and Legumes

Authors

N. R. Reddy, Ph.D.
Research Scientist
Department of Food Science and Technology
Virginia Polytechnic Institute and State University
Blacksburg, Virginia

M. D. Pierson, Ph.D.
Professor and Head
Department of Food Science and Technology
Virginia Polytechnic Institute and State University
Blacksburg, Virginia

S. K. Sathe, Ph.D.
Assistant Professor
Department of Nutrition and Food Science
Florida State University
Tallahassee, Florida

D. K. Salunkhe, Ph.D.
Professor Emeritus
Department of Nutrition and Food Science
Utah State University
Logan, Utah

CRC Press, Inc.
Boca Raton, Florida

Library of Congress Cataloging-in-Publication Data

Phytates in cereals and legumes / authors, N. R. Reddy ... [et al.].
      p. cm.
  Includes bibliographical references.
  ISBN 0-8493-6108-7
    1. Phytic acid -- Derivatives -- Physiological effect.   2. Phytic
acid -- Derivatives -- Metabolism.   3. Legumes -- Composition.   4. Grain -
- Composition.   I. Reddy, N. R.
    [DNLM: 1. Cereals -- analysis. 2. Legumes -- analysis. 3. Phytic
Acid -- analysis. QU 87 P578]
QP801.P634P47 1989
582.13′041924--dc20
DNLM/DLC
for Library of Congress   89-22175
                                                                                                                                           CIP

    This book represents information obtained from authentic and highly regarded sources. Reprinted material is quoted with permission, and sources are indicated. A wide variety of references are listed. Every reasonable effort has been made to give reliable data and information, but the author and the publisher cannot assume responsibility for the validity of all materials or for the consequences of their use.

    All rights reserved. This book, or any part thereof, may not be reproduced in any form without written consent from the publisher.

    Direct all inquiries to CRC Press, Inc., 2000 Corporate Blvd., N.W., Boca Raton, Florida, 33431

© 1989 by CRC Press, Inc.

International Standard Book Number 0-8493-6108-7

Library of Congress Card Number 88-24238
Printed in the United States

PREFACE

Cereals, legumes, and oilseed crops are grown over 90% of the world's total harvested area. These crops serve as a major source of nutrients for mankind. An important constituent of these foods is phytic acid. The salt form, phytate, commonly exists in cereals and legumes where it serves several physiological functions, especially during seed germination. Phytate is the major storage form of phosphorus and accounts for more than 80% of total phosphorus in cereals and legumes.

There is concern about the presence of phytate in cereals, legumes, and their derived foods since there is evidence that phytate decreases the bioavailability of essential minerals such as calcium, magnesium, iron, and zinc and proteins by forming complexes. Under normal physiological conditions phytate complexes are unavailable for absorption. On the other hand, there are potentially beneficial applications for cereal and legume phytates based on their interaction properties.

The purpose of this book is to bring together into one volume an overview of the extensive published data on cereal and legume phytates. The book consists of eleven chapters, in which the historical background; physiological functions and uses; biosynthesis and dephosphorylation; phytase enzyme; methods for analysis; occurrence, distribution, content, and dietary intake; interactions with minerals, proteins, carbohydrates, and enzymes; digestion and bioavailability; nutritional consequences; and technologies for removal of phytates from cereals and legumes are discussed in detail. The publication is intended as a reference for food scientists and technologists, especially those involved in phytate research and the technology and processing of cereals and legumes.

The authors would like to thank Drs. Munir Cheryan, Barbara F. Harland, Natholyn D. Harris, Frank A. Loewus, Mary W. Loewus, Eugene R. Morris, Gur S. Ranhotra, Bruce D. Rengers, and Connie M. Weaver, who critically reviewed various chapters of this book and made several constructive comments.

Special appreciation is extended to the individuals who so ably turned our handwriting into legible text: Ms. Sadie M. Freeman, Ms. Mary Jane Thompson, and the editorial staff of CRC Press, Inc.

N. R. Reddy
M. D. Pierson
S. K. Sathe
D. K. Salunkhe

AUTHORS

N. R. Reddy, Ph.D., is Research Scientist, Department of Food Science and Technology, Virginia Polytechnic Institute and State University, Blacksburg, Virginia.

Dr. Reddy obtained his B.Sc. (Agric.) and M.Sc. (Agric.) degrees from Andhra Pradesh Agricultural University, Hyderabad, India, in 1970 and 1972, respectively. He graduated in 1976 from Alabama A & M University, Normal, Alabama, with an M.S. in Food Science and received a Ph.D. degree in Nutrition and Food Science in 1981 from Utah State University, Logan, Utah.

Dr. Reddy is a member of the American Association of Cereal Chemists, the Institute of Food Technologists, and Sigma Xi. He has published more than 45 scientific papers, reviews, and book chapters and co-authored a book.

M. D. Pierson, Ph.D., is Professor and Head, Department of Food Science and Technology, Virginia Polytechnic Institute and State University (VPI & SU), Blacksburg, Virginia.

Dr. Pierson graduated in 1964 from Iowa State University, Ames, Iowa, with a B.S. degree in Biochemistry and obtained his M.S. and Ph.D. degrees in Food Science from the University of Illinois, Urbana, Illinois, in 1969 and 1970, respectively. He also served as a Research Chemist with George A. Hormel and Co., Austin, Minnesota.

Dr. Pierson is a fellow of the American Association for the Advancement of Science and a member of the American Society for Microbiology, the Society for Applied Bacteriology, American Meat Science Association, Institute of Food Technologists, International Association of Milk, Food, and Environmental Sanitarians, Gamma Sigma Delta, Phi Kappa Phi, Phi Tau Sigma, and Sigma Xi. He received the Gamma Sigma Delta VPI & SU Faculty Research Award in 1984 and served as chairman of the Institute of Food Technologists Food Microbiology Division.

He is a member of the editorial board of the *Journal of Food Safety* and served on the editorial boards of the *Journal of Food Science* and the *Journal of Food Protection*. Dr. Pierson served on numerous advisory panels and as a consultant to several food processing companies.

Under Dr. Pierson's guidance 30 graduate students received their M.S. or Ph.D. degrees. He has published more than 60 scientific papers, reviews, and book chapters and co-authored two books.

S. K. Sathe, Ph.D., is an Assistant Professor in the Department of Nutrition and Food Science, Florida State University, Tallahassee, Florida.

Dr. Sathe obtained his B.Sc. (Honors) in Chemistry in 1971, and received his B.Sc. (Tech.) and M.Sc. (Tech.) in Food Technology in 1974 and 1977, respectively, from Bombay University, Bombay, India. He graduated with a Ph.D. in Nutrition and Food Science from the Utah State University, Logan, Utah, in 1982.

Dr. Sathe is a member of the American Association of Cereal Chemists, the American Oil Chemists Society, the Institute of Food Technologists, Sigma Xi, and Phi Tau Sigma. He has published over 50 scientific papers, reviews, and book chapters.

D. K. Salunkhe, Ph.D., is Professor Emeritus, Department of Nutrition and Food Science, Utah State University, Logan, Utah.

Professor Salunkhe graduated with a B.Sc. (Agric.) degree with honors from Poona University, Poona, India, in 1949. In 1951 and 1954, respectively, he received his M.S. and Ph.D. degrees from Michigan State University, East Lansing, Michigan.

Professor Salunkhe was an Alexander Humbolt Senior Fellow and Guest Professor at the Technical University in Karlsruhe, West Germany. He was a Guest Lecturer at the Technological Institute, Moscow, U.S.S.R., and exchange scientist to Czechoslovakia, Rumania, and Bulgaria on behalf of the National Academy of Sciences, National Research Council, Washington, D.C., as well as advisor to the U.S. Army Food Research Laboratories and many food storage, processing, and consumer organizations.

He was Sigma Xi president, Utah State University chapter; Fellow of Utah Academy of Sciences, Arts, and Letters; Fellow of the Institute of Food Technologists; and Danforth Foundation Faculty Associate. He delivered the Utah State University's 50th Faculty Honor Lecture, "Food, Nutrition, and Health: Problems and Prospect" on the basis of his creative activities in research and graduate teaching.

Professor Salunkhe served on the editorial boards of the *Journal of Food Quality, International Journal of Qualitas Plantarum/Plant Foods for Human Nutrition,* the *Journal of Food Science,* and the *Journal of Food Biochemistry.*

Under Professor Salunkhe's guidance 85 graduate students received their M.S. or Ph.D. degrees. He has authored more than 400 scientific papers, reviews, book chapters, and reports. Some of his papers have received recognition and awards as outstanding articles in biological journals. He has authored and/or co-authored 25 books.

TABLE OF CONTENTS

Chapter 1
Introduction: Historical Background, Chemical Structure, and Properties of Phytic Acid 1
    References 5

Chapter 2
Biosynthesis and Dephosphorylation of Phytates 7
    I.   Introduction 7
    II.   Biosynthesis of Phytates 7
    III. Dephosphorylation of Phytates 10
    References 12

Chapter 3
Phytase Enzyme: Biosynthesis and Characterization 15
    I.   Introduction 15
    II.   Phytase Synthesis and Location in Seeds 15
    III. Phytase Extraction and Purification 16
    IV. Molecular Properties of Phytases 16
    References 19

Chapter 4
Physiological Functions and Uses of Phytates 23
    I.   Introduction 23
    II.   Physiological Functions 23
    III. Commercial Manufacture and Uses 24
    References 25

Chapter 5
Methods for Analysis of Phytate 27
    I.   Introduction 27
    II.   Qualitative Methods 27
        A.   Paper Chromatography 27
        B.   Paper Electrophoresis 27
        C.   Thin-Layer Chromatography 28
        D.   Ion Exchange Chromatography 28
    III. Quantitative Methods 28
        A.   Precipitation Methods 29
            1.   Indirect Methods 29
            2.   Direct Methods 31
        B.   Chromatographic Methods 31
            1.   Ion Exchange Chromatographic Methods 31
            2.   High Performance Liquid Chromatographic Methods 33
        C.   Nuclear Magnetic Resonance and Other Methods 35
    References 36

Chapter 6
Occurrence, Distribution, Content, and Dietary Intake of Phytate 39
    I.   Introduction 39
    II.   Occurrence 39

	III.	Phytate Distribution and Content	41
		A. Phytate Distribution	42
		B. Phytate Content	42
		1. Cereals and Cereal Products	42
		2. Legumes and Legume Products	46
		C. Effects of Environmental and Other Factors on the Phytate Content of Cereals and Legumes	47
	IV.	Dietary Intake of Phytate	47
	References		50

Chapter 7
Interactions of Phytate with Proteins and Minerals ... 57

I.	Introduction		57
II.	Phytate-Protein Interactions		57
	A.	Formation of Phytate-Protein Complexes	57
		1. Low pH (3.5 or Less)	57
		2. Intermediate pH	58
		3. High pH	59
	B.	Interactions of Phytate with Soy Proteins	59
	C.	Interactions of Phytate with Other Proteins	61
	D.	Effects of Phytate-Protein Interactions on Protein Functionality	62
III.	Phytate-Mineral Interactions		66
References			69

Chapter 8
Phytate Digestion and Bioavailability ... 71

I.	Introduction	71
II.	Animals	71
III.	Poultry	72
IV.	Rats	74
V.	Humans	75
References		76

Chapter 9
Nutritional Consequences of Phytates ... 81

I.	Introduction		81
II.	Minerals		81
	A.	Zinc	81
		1. Animal Studies	82
		2. Human Studies	86
		3. Phytate/Zinc Molar Ratio for Predicting Zinc Bioavailability	88
	B.	Iron	92
		1. Animal Studies	92
		2. Human Studies	95
	C.	Calcium	96
		1. Animal Studies	96
		2. Human Studies	96
	D.	Magnesium	98
		1. Animal Studies	98
		2. Human Studies	99
	E.	Other Minerals	99

III.	Enzyme Inhibition 100
IV.	Starch and Protein Digestibility 100
	References 102

Chapter 10
Influence of Processing Technologies on Phytate 111
I.	Introduction 111
II.	Cooking 111
III.	Germination 115
IV.	Fermentation and Breadmaking 120
V.	Soaking, Autolysis, and Other Processes 126
	References 132

Chapter 11
Technology of Phytate Removal 137
I.	Introduction 137
II.	Mechanical Processes 137
III.	Selective Extraction and Differential Solubility Methods 137
IV.	Membrane Separation 141
V.	Ion Exchange Process 143
VI.	Enzyme Treatment 145
VII.	Genetic Selection 145
	References 146

Index 149

Chapter 1

INTRODUCTION: HISTORICAL BACKGROUND, CHEMICAL STRUCTURE, AND PROPERTIES OF PHYTIC ACID

Cereals and legumes contain significant amounts of phosphorus in the form of phytic acid (myoinositol hexaphosphate). It serves several physiological functions and also significantly influences the functional and nutritional properties of cereals and legumes and their derived foods by complexing with proteins and essential minerals. The terms phytic acid, phytate, and phytin refer to free acid, salt, and calcium/magnesium salt, respectively. In the literature, the name phytic acid has been used interchangeably with the term phytate, which is a salt.

The discovery and early research on the isolation, chemical structure, and properties of phytate have been reviewed by Rose,[1] Mellanby,[2] and Reddy, et al.[3] Our knowledge of phytate has its beginning in the discovery by Hartig,[4,5] who isolated small, nonstarch grains from various plant seeds. He considered these grains to be an essential reserve material that was important in the seed germination and plant growth. Later, Pfeffer[6] differentiated these small grains into three groups: (1) crystals of calcium oxalate, (2) a protein substance, and (3) a compound giving no reactions for protein, fat, or inorganic salts. The third group was found in all of the 100 different seeds which he examined. Pfeffer characterized the third group as having rounded surfaces, assuming spheroidal shapes, and frequently twinning so as to present a convoluted appearance; they were free of nitrogen, but contained calcium, magnesium, and phosphorus. He named this third group of grains *globoids*. Organic matter was also noted in the globoids and the suggestion was made that the substance was a phosphate combined with a carbohydrate. While studying the proteins of Indian mustard, Palladin[7] obtained from the fat-free, finely ground seeds a substance which was soluble in 10% sodium chloride, but precipitated on heating. This substance was soluble in cold and insoluble in hot water. By filtering off the permanent coagulum, reheating the filtrate, and filtering while hot, he obtained a fairly pure product that was rich in phosphorus and contained calcium and magnesium, but no nitrogen. Palladin's work was later confirmed by Schulze and Winterstein,[8] who felt that the compound obtained by this isolation procedure was identical with Pfeffer's globoid particles. Subsequently, Winterstein[9] suggested "inositephosphoric acid" as the proper name for this compound, since it yielded inosite and phosphoric acid on hydrolysis.

Posternak[10-12] extensively studied this substance. In the early stages of his work, he rejected the name suggested by Winterstein and proposed a structural formula (Figure 1), which did not include the inosite (inositol) ring. He named the substance *phytin* (derived from Greek) and under this trade name it has long been marketed by a chemical company in Basel, Switzerland. Based on more detailed chemical tests, Posternak proposed a different structural formula $C_2H_8O_9P_2$ (Figure 2). He believed that inosite was a hydrolysis product, formed when phytin was heated under pressure with mineral acids. However, early researchers expressed doubt about inosite being formed as a hydrolysis product of phytin.[1] Suzuki et al.[13] obtained inosite from phytin by the action of a rice bran enzyme. They concluded that phytin was the hexaphosphoric acid ester of inosite and proposed a structure, $C_6H_{18}O_{24}P_6$ (Figure 3), for phytin. Neuberg[14] also concluded that inosite was a component of phytin. He obtained inosite and furfurol on mixing phytin with phosphoric acid and distilling under reduced pressure. He also showed that furfurol can be obtained from inosite. Based on his studies, Neuberg[14] proposed a slightly different structure, $C_6H_{24}O_{27}P_6$ (Figure 4), for phytin that can be distinguished mainly by having three P–O–P linkages between pairs of adjacent phosphates.

Levene,[15] as a result of studies on hempseed grain, concluded that phytin contains phosphate, inosite, and a carbohydrate of the pentose group. His work was later criticized by Neuberg,[16] who claimed that there were impurities in the preparation. Starkenstein[17,18] declined to accept the

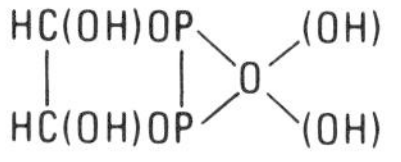

FIGURE 1. Structural formula for phytin first proposed by Posternak,[10,11] $C_2H_6O_7P_2$.

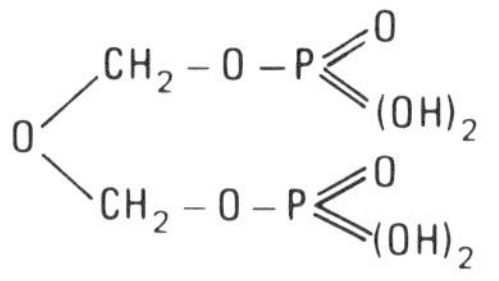

FIGURE 2. Second structural formula for phytin proposed by Posternak.[12] Anhydrohydroxymethylene disphosphoric acid, $C_2H_8O_9P_2$.

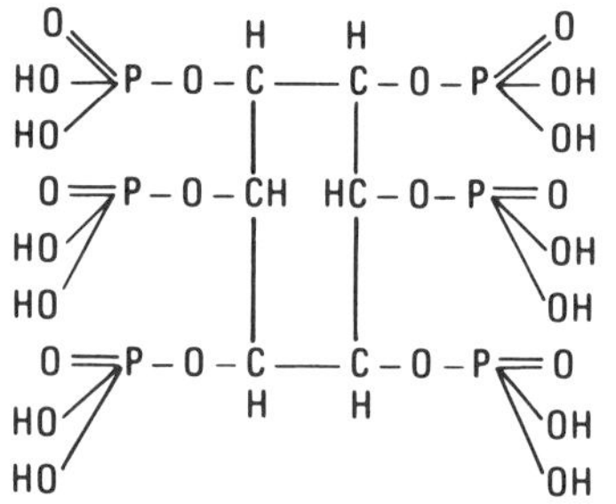

FIGURE 3. Structural formula for "phytin" proposed by Suzuki et al.,[13] $C_6H_{18}O_{24}P_6$.

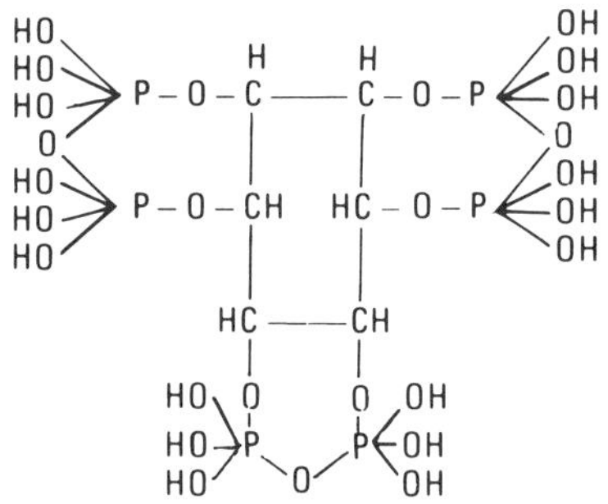

FIGURE 4. Structure of "phytin" as proposed by Neuberg,[14] $C_6H_{18}O_{27}P_6$.

simple structure proposed for phytin by Posternak; therefore, he proposed a different structure, $C_6H_{24}O_{27}P_6$ (Figure 5). He contended that the phosphoric acid was in the pyro-form. This structure for phytin corresponds to a hexaphosphoric acid ester of inosite plus $3H_2O$. Anderson[20,21] prepared inosite phosphoric acid salts from various sources and determined the compositions that correspond to inositehexaphosphoric acid structure, $C_6H_{18}O_{24}P_6$ or $C_6H_6O_6[PO(OH)_2]_6$ (Figure 6). Within a few years, it became generally acceptable that phytic acid was the hexaphosphate of myoinositol.[22-25]

There are nine stereoisomers of inositol, of which seven are meso structures and two form a chiral pair. They are (1) *cis*-, (2) *epi*-, (3) *allo*-, (4) *neo*-, (5) *myo*-, (6) *muco*-, (7) 1L-*chiro*-, (8) 1D-*chiro*-, and (9) *scyllo*-inositol.[26] Of these, seven occur in nature either as free or combined.

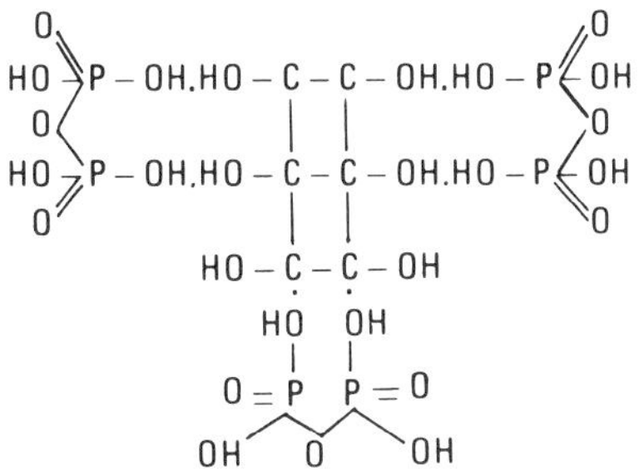

FIGURE 5. Structure for "phytin" as proposed by Starkenstein,[19] $C_6H_{24}O_{27}P_6$.

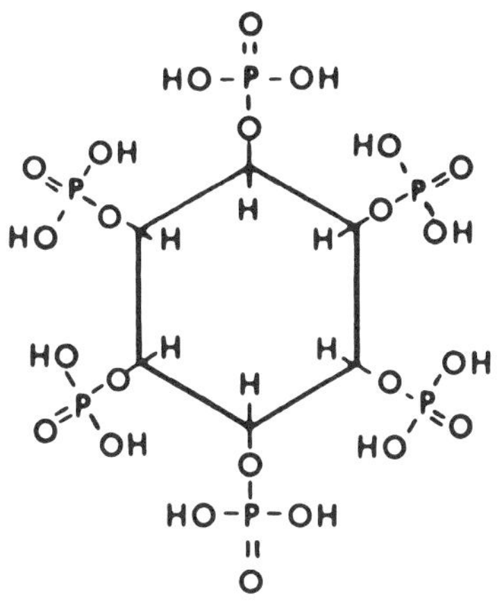

FIGURE 6. Structure for phytic acid as proposed by Anderson,[21,22] $C_6H_{18}O_{24}P_6$.

The exceptions are *epi-* and *allo-*inositol. The myoinositol is common in plants. The *chiro-*, *scyllo-*, and *neo-*inositol hexaphosphates have been isolated from soils.[27-29]

The structure of phytic acid had been the subject of controversy. For many years the controversy was centered on the structure proposed by Anderson and the structure suggested by Neuberg (Figure 7). Several additional structures for phytic acid[25,28,30-37] have also been proposed. The primary point of contention has been the isomeric conformation of the phosphate groups within the compound and whether or not three strongly bound water molecules were incorporated into the structure.

The data of some studies appeared to support the Neuberg structure.[24,32,35,38] Posternak[24] reported that the crystalline salt of phytic acid retained three molecules of water even after prolonged drying at 100°C. He indicated the formula of phytic acid as $C_6H_{18}O_{24}P_6 \cdot 3H_2O$ but did not suggest how the strongly bound water molecules were incorporated into the phytic acid structure. This observation seems to favor the Neuberg structure. Gosselin and Coghlan[32] in their studies on calcium-phytic acid interactions concluded that there must be P–O–P linkages between pairs of adjacent phosphates within the phytic acid molecule. Brown et al.[35] found that only 12 acid hydrogens were titratable in aqueous solution and 18 hydrogens could be detected in glacial acetic acid solution. Six of the 18 hydrogens were too weakly acidic to be titratable in water. Furthermore, elemental analysis, titrations of sodium phytate with metal ions, and titration of phytic acid solutions containing an excess of metal ions all favored the 18 acid-hydrogen structure for phytic acid proposed by Neuberg.

Other studies appeared to favor the Anderson structure.[33,34,36,39] From a chemical hydrolysis study, Desjobert and Fleurent[34] concluded that phytic acid behaved consistently with the

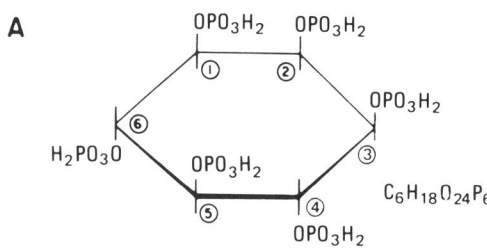

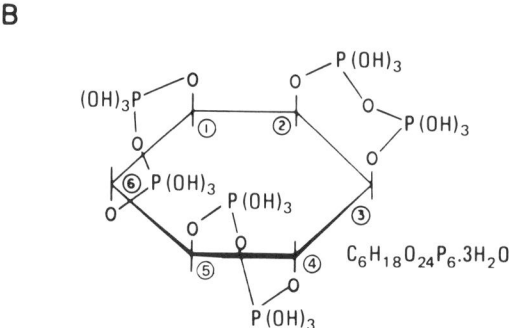

FIGURE 7. Proposed structures for phytic acid. Structure A was suggested by Anderson[20,21] and structure B by Neuberg.[14]

Anderson model. The research of Barré et al.[33] on pH-titration and conductivity curves also supported Anderson's structure. In Posternak's review[25] it is suggested that only 12 dissociable hydrogens per molecule can be detected by potentiometric titration in aqueous solution; this has been confirmed by Maddaiah et al.[40] This further supports the Anderson structure for phytic acid. Also, studies using ^{31}P nuclear magnetic resonance (NMR)[36,39] and X-ray crystallography[37,41] have left little doubt that the structure proposed by Anderson was, in fact, the predominant form found in plant seeds and/or grains. The nomenclature for inositol phosphates has been revised.[42] The new nomenclature for phytic acid in plant seeds is myoinositol-1, 2, 3, 5/4, 6-hexakis (dihydrogen phosphate).

The conformational structures for phytate have been derived from X-ray analysis,[37] ^{31}P NMR and pH titrations,[39] and ^{31}P NMR.[36] However, there is some controversy about the conformation of the phytate molecule. Johnson and Tate[36] suggested that the phosphate at C-2 position is in the axial position, while the phosphate groups on C-1, C-3, C-4, C-5, and C-6 are equatorial. On the other hand, Blank et al.,[37] through a single crystal X-ray analysis, concluded that the phosphate groups at positions C-1, C-3, C-4, C-5, and C-6 are axially disposed with that at C-2 position equatorial rather than the reverse indicated by Johnson and Tate.[36] Data of Costello et al.[39] appear to support the conformation suggested by Johnson and Tate.[36] Isbrandt and Oertel[43] resolved a long-standing question of the solution conformation of myoinositol hexaphosphate through combined use of ^{13}C NMR, ^{31}P NMR, and Raman Spectroscopy. Myoinositol hexaphosphate exists in either of two conformations in aqueous solution depending on pH: under acidic conditions 1-axial/5-equatorial conformer predominates, while under strong alkaline conditions, the inverted conformer, i.e., 5-axial/1-equatorial, prevails. At pH 9.40 and

27°C, a 0.1 M sodium phytate solution contains equal amounts of two conformers.[43]

Dissociation constants for myoinositol hexaphosphates were determined by Barré et al.[33] from potentiometric titrations. They demonstrated that, of the twelve replaceable protons in the phytic acid molecule, six are strongly dissociated with a pKa of about 1.84, two are weak acid functions with a pKa of 6.30, and four are weakly dissociated at a pKa of 9.70, and that they can not be determined by usual titration methods. Using ^{31}P NMR and pH titration methods, Costello et al.[39] determined pKa values for dissociating groups of phytic acid. They obtained results similar to that of the potentiometric titration method: six in the strong acid range (pKa 1.1 to 2.1), one in the weak acid range (pKa 5.70), two with pKa 6.80 to 7.60, and three in the very weak acid range (pKa 10.0 to 12.0). This suggests that the phytic acid has a tremendous potential for complexing positively charged proteins or multivalent cations in many foods, since it exists as a strong negatively charged molecule over a wide range of pH values. The solubility characteristics of phytates have been extensively reviewed and discussed.[24]

Phytate accounts for up to 85% of the total phosphorus in many cereals and legumes. Concern about the presence of phytate in cereals and legumes and their derived food products arises from the evidence that it decreases the bioavailability of essential minerals by forming complexes. Many of the phytate-mineral complexes are insoluble and may be unavailable for absorption under normal physiological conditions. This interference with intestinal absorption of minerals may lead to mineral deficiencies[45] in humans and animals. The presence of phytate also influences the functional and nutritional properties of proteins.[44] Occurrence, distribution and content, dietary intake, nutritional consequences, and technology for removal of phytate from cereals and legumes are discussed in the subsequent chapters. The chemistry and interactions of phytate with other food components and enzymes; biological functions, beneficial effects, and applications of phytate; and methods for determination of phytate are also presented in the subsequent chapters of this book.

REFERENCES

1. **Rose, A. R.,** A resumé of the literature on inosite phosphoric acid, with special reference to the relation of that substance to plants, *Biochem. Bull.,* 2, 32, 1912.
2. **Mellanby, E.,** *A Story of Nutritional Research,* Williams and Wilkins, Baltimore, MD, 1950, 248.
3. **Reddy, N. R., Sathe, S. K., and Salunkhe, D. K.,** Phytates in legumes and cereals, *Adv. Food Res.,* 28, 1, 1982.
4. **Hartig, T.,** Uber das Klebermehl, *Bot. Ztg.,* 13, 881, 1855.
5. **Hartig, T.,** Weitere Mittheilunger über Klebermehl, *Bot. Ztg.,* 14, 257, 1856.
6. **Pfeffer, W.,** Untersuchungen über die Proteinkorner und die Bedeu des Asparagin beim Keimen der Samen, *Jahrb. Wiss. Bot.,* 8, 429, 1872.
7. **Palladin, W.,** *Z. Biol.,* 31, 191, 1894 (cited from **Rose, A. R.,** *Biochem. Bull.,* 2, 21, 1912).
8. **Schulze, E. and Winterstein, E.,** *Z. Physiol. Chem.,* 40, 120, 1896 (cited from **Rose, A. R.,** *Biochem. Bull.,* 2, 32, 1912).
9. **Winterstein, E.,** *Ber. Dtsch. Chem. Gess.,* 30, 2299, 1897 (cited from **Rose, A. R.,** *Biochem. Bull.,* 2, 21, 1912).
10. **Posternak, S.,** Sur la constitution de l'acide phosphorganique de reserve des plantes vertes et sur le premier produit de reduction du gaz carbonizue dans l'acte, *C.R. Acad. Sci.,* 137, 439, 1903.
11. **Posternak, S.,** Sur un nouveau principe phospho-organique d'arigine végétale la phytine, *C.R. Soc. Biol.,* 55, 1190, 1903.
12. **Posternak, S.,** Sur la composition chimique et la signification des grains d'aleurone, *C.R. Acad. Sci.,* 140, 322, 1905.
13. **Suzuki, U., Yoshimura, K., and Takaishi, M.,** Uber ein enzym "phytase" das anhydro-oxy-methylen disphosphorsaure Spaltet, *Bull. Coll. Agric. Tokyo Imp. Univ.,* 7, 495, 1907.

14. **Neuberg, C.,** Beziehung des cyclischen inosits zu den aliphatischen zuckern, *Biochem Z.,* 9, 551, 1908.
15. **Levene, P. A.,** The conjugated phosphoric acid in plant seeds, *Biochem. Z.,* 16, 399, 1909.
16. **Neuberg, C.,** Note concerning phytin, *Biochem. Z.,* 16, 406, 1909.
17. **Starkenstein, E.,** Inosituria and the physiological significance of inosite, *Z. Exp. Pathol Ther.,* 5, 378, 1908.
18. **Starkenstein, E.,** The biological importance of inosite phosphoric acid, *Biochem. Z.,* 30, 56, 1910.
19. **Starkenstein, E.,** Ion action of phosphoric acids, *Biochem. Z.,* 32, 243, 1911.
20. **Anderson, R. J.,** Phytin and phosphoric acid esters of inosite, *J. Biol. Chem.,* 11, 471, 1912.
21. **Anderson, R. J.,** Phytin and phosphoric acid esters of inosite, *J. Biol. Chem.,* 12, 97, 1912.
22. **Anderson, R. J.,** Synthesis of phytic acid, *J. Biol. Chem.,* 43, 117, 1920.
23. **Starkenstein, E.,** Pharmacological action of acids which precipitate calcium and of magnesium salts, *Arch. Exp. Pathol. Pharmakol.,* 77, 45, 1914.
24. **Posternak, S.,** The synthesis of inositohexaphosphoric acid, *Helv. Chim. Acta,* 4, 150, 1921.
25. **Posternak, T.,** *Les Cyclitols: Chimie, Biochimie, Biologie,* Hermann, Paris, 1962, 223.
26. **Loewus, F. A. and Loewus, M. W.,** Myo-inositol: biosynthesis and metabolism, in *The Biochemistry of Plants,* Preiss, J., Ed., Academic Press, New York, 1980, 43.
27. **Cosgrove, D. J.,** The chemistry and biochemistry of inositol polyphosphates, *Rev. Pure Appl. Chem.,* 16, 209, 1966.
28. **Cosgrove, D. J.,** *Inositol phosphates: Their Chemistry, Biochemistry, and Physiology,* Elsevier, New York, 1980, 26.
29. **Smith, D. H. and Clark, F. E.,** Anion-exchange chromatography of inositol phosphates from soil, *Soil Sci.,* 72, 353, 1951.
30. **Posternak, S. and Posternak, T.,** Sur la configuration de l'inosite inactive, *Helv. Chim. Acta.,* 12, 1165, 1929.
31. **Wrenshall, C. L. and Dyer, W. J.,** Organic phosphorus in soil. II. The nature of the organic phosphorus compounds, nucleic acid derivatives and phytin, *Soil Sci.,* 51, 235, 1941.
32. **Gosselin, R. E. and Coghlan, E. R.,** The stability of complexes between calcium and orthophosphate, polymeric phosphate, and phytate, *Arch. Biochem. Biophys.,* 45, 301, 1953.
33. **Barré, R., Curtois, J. E., and Wormser, G.,** Étude de la structure de l'acide phytique au moyen de ses courbes de titration et de la conductivité de ses solutions, *Bull. Soc. Chem. Biol.,* 36, 455, 1954.
34. **Desjobert, A. and Fleurent, P.,** Influence de la réaction du miléen sur l'hydrolyse chimique de l'inositol-héxaphosphate: considerations sur la constitution de ce dérive, *Bull. Soc. Chem. Biol.,* 36, 475, 1954.
35. **Brown, E. C., Heit, M. L., and Ryan, D. E.,** Phytic acid: an analytical investigation, *Can. J. Chem.,* 39, 1290, 1961.
36. **Johnson, L. F. and Tate, M. E.,** Structure of phytic acids, *Can. J. Chem.,* 47, 63, 1969.
37. **Blank, G. E., Pletcher, J., and Sax, M.,** The structure of myo-inositol hexaphosphate, dodecasodium salt octama-contahydrate: a single crystal x-ray analysis, *Biochem. Biophys. Res. Commun.,* 44, 319, 1971.
38. **Fischler, F. and Kurtern, F. H.,** Uber einfachere Nanchweismethoden von Inosit und Phytinaten sowie über ein definertes, kristallisertes, Barium-phytinat, *Biochem. Z.,* 254, 138, 1932.
39. **Costello, A. J. R., Glonek, T., and Myers, T. C.,** Phosphorus-31 nuclear magnetic resonance — pH titrations of myoinositol hexaphosphate, *Carbohydr. Res.,* 46, 159, 1976.
40. **Maddaiah, V. T., Kurnick, A. A., and Reid, B. L.,** Phytic acid studies, *Proc. Soc. Exp. Biol. Med.,* 115, 391, 1964.
41. **Truter, M. A. and Tage, M. E.,** Crystallographic studies of hydrates of dodecasodium myo-inositol hexaphosphate (phytic acid), *J. Chem. Soc.(B),* 70, 40, 1970.
42. **IUPAC-IUB,** The nomenclature of cyclitols, *Eur. J. Biochem.,* 5, 1, 1968.
43. **Isbrandt, L. R. and Oertel, R. P.,** Conformational states of myo-inositol hexakis (phosphate) in aqueous solution A ^{13}C NMR, ^{31}P NMR, and Raman Spectroscopic Investigation, *J. Am. Chem. Soc.,* 102, 3144, 1980.
44. **Cheryan, M.,** Phytic acid interactions in food systems, *CRC Crit. Rev. Food Sci. Nutr.,* 13, 297, 1980.
45. **Graf, E.,** Applications of phytic acid, *J. Am. Oil Chem. Soc.,* 60, 1861, 1983.

Chapter 2

BIOSYNTHESIS AND DEPHOSPHORYLATION OF PHYTATES

I. INTRODUCTION

The biosynthesis, accumulation, and dephosphorylation of phytate in seeds and grains[1-14] is thought to be confined to electron-dense regions called globoids or aleurone particles, although recent observations of Greenwood and Bewley[61] and Organ et al.[62] report on biosynthesis and accumulation of phytate-containing particles in cytoplasm prior to globoid formation. In seeds and/or grains, the globoids are associated with germ, aleurone layer, scutellum, and cotyledons or endosperm, according to the species involved.[14,15] For instance, in cereals, with some exceptions, phytate is abundant in the aleurone layer. In corn, over 80% of phytate is present in the germ. Large amounts of phytate are found in cotyledons of dicotyledonous seeds including legumes. During seed maturation, phytate rapidly accumulates in globoids. The enzymes and steps involved in the biosynthesis, accumulation, and dephosphorylation of phytate in seeds are known only in part. The extensive literature on the biosynthesis, accumulation, and dephosphorylation of phytate has been the subject of several reviews.[11,15-22]

II. BIOSYNTHESIS OF PHYTATES

Cosgrove[17] suggested three possible mechanisms for the biosynthesis of phytate. These include (1) phosphorylation of phosphoinositide intermediates and subsequent hydrolysis to generate corresponding inositol phosphates, (2) successive phosphorylation in the absence of free intermediates at each step, and (3) direct stepwise phosphorylation of free myoinositol and/or myoinositol monophosphate by a kinase type of reaction. The first two mechanisms for phytate formation have received either little or no support. The literature related to these two mechanisms has been reviewed.[19] Most studies favor a direct stepwise phosphorylation pathway from myoinositol or myoinositol monophosphate to the myoinositol hexakisphosphate.[1,3,23-26]

Many researchers have studied the biosynthesis of phytate in whole plant, plant organ, subcellular organelles or cell culture, and cell-free extracts using radioisotopic substrates.[15] Some of the approaches used include injection of labeled myoinositol into immature pea pods,[27] soaking of ripening heads of rice in labeled inorganic phosphate or myoinositol,[16,23,28] incubation of globoids or aleurone particles from sunflower,[1] and rice grains[8] with labeled inorganic phosphate or myoinositol, addition of labeled phosphorus or myoinositol to media containing cultured rice cells,[29-31] uptake of labeled glucose, myoinositol, or inorganic phosphate by various duckweeds,[32,33] imbibitory uptake of labeled inorganic phosphate or myoinositol by mung bean seeds[24] and wheat,[34] injection of labeled myoinositol into peduncles of ripening wheat,[12] and cell-free studies involving germinating mung bean extracts with labeled myoinositol and myoinositol phosphates.[35-37] Most of the above and other studies[25,26,38] support the view that the free myoinositol and glucose-6-phosphate play an important role in the formation of phytate. Glucose-6-phosphate is a precursor for phytate biosynthesis (Figure 1) and is catalyzed to 1L-myoinositol-1-phosphate by a NAD⁺ dependent enzyme, 1L-myoinositol-1-phosphate synthase, during seed development.[22] This enzyme has been isolated and purified from a number of plant species.[20] De and Biswas[38] proposed a direct role for synthase-produced 1L-myoinositol-1-phosphate in phytate biosynthesis; however, the experimental evidence is lacking. Based on their work with *in vivo* labeling studies and identification of an *in vitro* product of phosphorylation of myoinsositol as myoinositol-2-phosphate, others[29,30,39] have suggested myoinositol-2-phosphate as the myoinositol monophosphate precursor of phytate. However, Loewus[22] suggested that these conclusions should be treated with caution in view of potential acid-

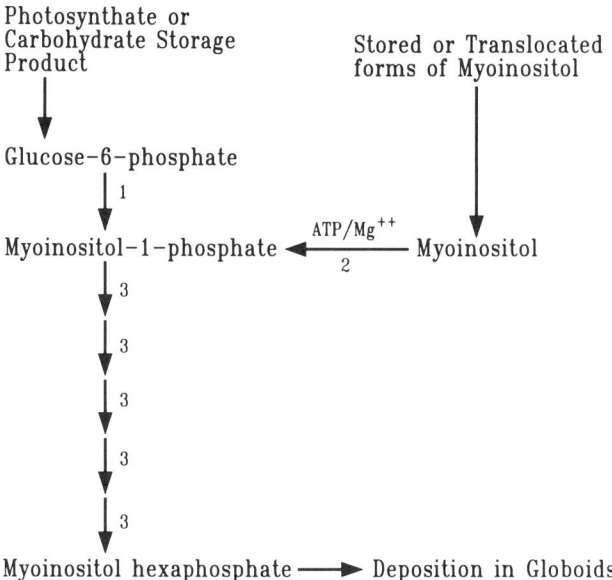

FIGURE 1. Schematic diagram for biosynthesis of phytate in seeds.[19,20,23]

catalyzed migration of phosphate from C-1 or C-3 of myoinositol to C-2[18] or involvement of an intermediary cyclic myoinositol-1,2-monophosphate. In plants, several salvage pathways contribute to free myoinositol, in addition to *de novo* biosynthesis of myoinositol via 1L-myoinositol-1-phosphate synthase and through hydrolysis of 1L-myoinositol-1-phosphate to free myoinositol by an alkaline, Mg^{2+}-dependent enzyme, myoinositol-1-phosphatase.[15,22] Free myoinositol participates in many metabolic processes, notably myoinositol oxidation, galactinol biosynthesis, phosphatidylinositol biosynthesis, and the formation of isomeric inositols and their esters,[21,22] and it also serves as an initial substrate for phytate biosynthesis (Figure 1). Myoinositol kinase (an ATP-specific enzyme) is widely distributed in nature including plants[15,22] and phosphorylates free myoinositol to 1L-myoinositol-1-phosphate.[40] In this respect, myoinositol kinase provides a salvage mechanism for returning free myoinositol into the same stereoisomeric form of 1L-myoinositol-1-phosphate as that produced by 1L-myoinositol-1-phosphate synthase. If phosphorylation of 1L-myoinositol-1-phosphate rather than myoinositol-2-phosphate is regarded as the first committed step in phytate biosynthesis, then two mechanisms exist for supplying precursor, the 1L-myoinositol-1-phosphate synthase which depends on hexose phosphate (glucose-6-phosphate) production and the myoinositol kinase which is dependent on a renewable source of free myoinositol.[22] There is a need for further studies on the regulatory events surrounding 1L-myoinositol-1-phosphate biosynthesis.

The stepwise phosphorylation of myoinositol-1-phosphate to myoinositol hexakisphosphate (phytate) by kinase enzymes has been proposed.[25,26,38] However, questions remain about the appearance of discrete intermediary phosphorylated forms of myoinositol between myoinositol monophosphate and phytate.[15] Biswas and co-workers,[35-37] working with germinating mung bean seeds, partially purified and characterized phosphoinositol kinase that catalyzes successive phosphorylation of myoinositol-1-phosphate and/or myoinositol-2-phosphate. Phosphoinositol

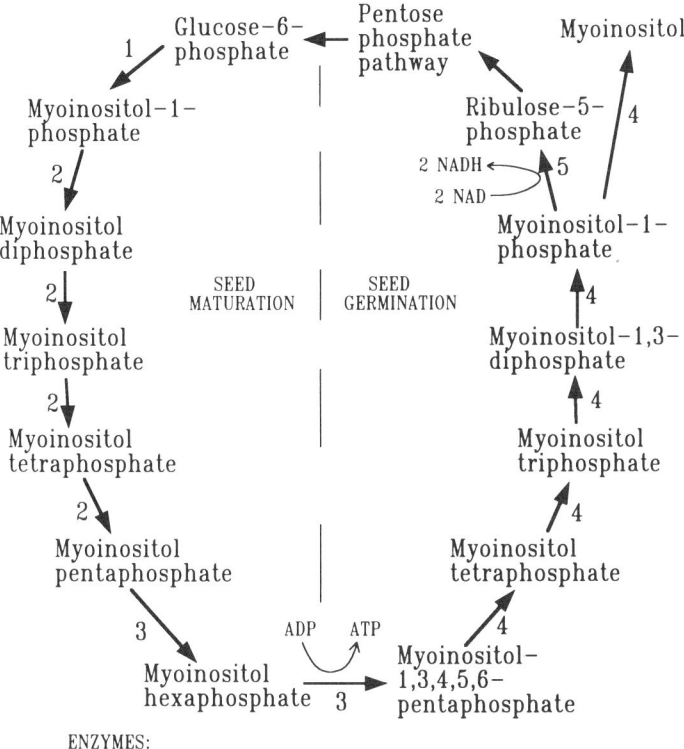

FIGURE 2. Proposed metabolic cycle involving myoinositol phosphates during maturation and germination of seeds.[26,38]

kinase phosphorylates myoinositol-1-phosphate stepwise to myoinositol pentaphosphate when myoinositol-1-phosphate is used as a substrate.[37] It also phosphorylates stepwise up to myoinositol hexakisphosphate if myoinositol-2-phosphate is used as the substrate. Biswas et al.[26,41] and De and Biswas[38] suggested that the action of a reversible phytate-ADP-phosphotransferase is required for complete phosphorylation, starting from myoinositol-1-phosphate (Figure 2). Phytate-ADP-phosphotransferase is responsible for synthesis of myoinositol hexakisphosphate from myoinositol pentaphosphate. This enzyme has been isolated, purified, and characterized from germinating mung bean seeds.[41] This mechanism has not been examined in other phytate synthesizing legumes or cereal grains.

Igaue et al.[29,30] studied suspension-cultured rice cells and found further evidence for stepwise phosphorylation of myoinositol to produce phytate. They found decreased amounts of individual myoinositol phosphates in the order of their phosphorylation, except for myoinositol monophosphate and myoinositol hexakisphosphate. Igaue et al.[30] identified and characterized the structures of myoinositol phosphate-isomers in the cultured rice cells by proton-decoupled ^{31}P nuclear magnetic resonance (NMR). Their[30] results suggested two possible biosynthetic pathways for phytate, one beginning with myoinositol-1-phosphate and the other with myoinositol-2-phosphate. The intermediate phosphorylated forms identified were myoinositol-1,3-diphosphate, myoinositol-1,3,5-triphosphate, myoinositol-1,2,3,4,5-tetraphosphate, and myoinositol-1,3,4,5,6-pentaphosphate for the pathway beginning with myoinositol-1-phosphate, and myoinositol-2,4-diphosphate, myoinositol-2,4,5-triphosphate, myoinositol-1,2,4,5-tetraphosphate and myoinositol-1,2,4,5,6-pentaphosphate(I) and myoinositol-1,2,3,4,5-pentaphosphate for the pathway starting with myoinositol-2-phosphate. Some of these intermediates, myoinositol-1,3,4,5,6-pentaphosphate, myoinositol-1,2,3,4,5,-pentaphosphate, and myoinositol-1,3-diphosphate are reported to have structures in common with those of phytase hydrolysis products.[18]

The stereoisomeric form of myoinositol monophosphate required for phytate biosynthesis still needs to be conclusively determined, although most evidence points toward 1L-myoinositol-1-phosphate.[15] In summary, it appears that the biosynthesis of phytate in plants and developing seeds occurs via two possible pathways, one beginning with myoinositol-1-phosphate and the other with myoinositol-2-phosphate as initial substrates. In both cases, phosphorylation proceeds stepwise at least up to myoinositol pentaphosphate if myoinositol-1-phosphate is available as initial substrate and possibly to phytate if myoinositol-2-phosphate is available as initial substrate. One or more kinase type of enzymes catalyze the phosphorylation steps. In the case of the first pathway, there is a stepwise phosphorylation by phosphoinositol kinase to produce myoinositol pentaphosphate and an additional phytate-ADP-phosphotransferase completes the sequence by adding a sixth phosphate to yield myoinositol hexakisphosphate. A single enzyme, phosphoinositol kinase, phosphorylates stepwise starting with myoinositol-2-phosphate to produce myoinositol hexakisphosphate in the second pathway.

III. DEPHOSPHORYLATION OF PHYTATES

In vivo metabolism of inositol hexaphosphate has not been described.[60] However, *in vivo* phytate metabolism can be deduced from experiments with [2-^{14}C] myoinositol-labeled duckweed where all six phosphate esters were detected.[32] The duckweed synthesized labeled phytic acid from labeled myoinositol and subsequent breakdown led to intermediate phosphate esters. *In vitro*, a complete phosphorylation of phytate by chemical and enzymatic means proceeds by six successive steps yielding five classes of intermediates.[17,18,58] These intermediates can be easily isolated from the reaction mixtures. During germination of cereal grains and legumes, dephosphorylation of phytate occurs in a stepwise way. It is catalyzed by one or more phytases resulting in release of inorganic phosphate and myoinositol to meet the biosynthetic needs in the growing tissues.[15,22] The phytases from various sources (plants, fungi, and bacteria) exhibit distinctive stereospecificities towards the dephosphorylation of phytate.[18] For example, on hydrolysis of phytate by phytases from plant sources, the initial product is L-myoinositol 1,2,3,4,5-pentaphosphate (Figure 3). On the other hand, phytases from microbial sources hydrolyze phytate and yield as an initial product D-myoinositol 1,2,4,5,6-pentaphosphate. Dephosphorylation of phytate by phytases isolated from wheat bran and germinated mung beans have been extensively investigated.[42-46] Wheat bran contains two phytases, F1(6-phytase) and F2(2-phytase), of which F1 is responsive to lipid activation.[43] Dephosphorylation of phytate by F1 phytase from wheat bran yields as a first product 1L-myoinositol-1,2,3,4,5-pentaphosphate and subsequently removes phosphate from carbons 5, 4, 3 (or 1), and 1 (or 3) to produce myoinositol-2-phosphate (Figure 3), the ultimate product. The F2 phytase dephosphorylates phytate to produce myoinositol-1-phosphate as the final product, possibly as the 1L isomer, although this remains to be proven. Mung bean phytase gives a different pattern of hydrolysis products.[45] Dephosphorylation of phytate by mung bean phytase liberates phosphate from carbon 6 to yield as a first product, myoinositol-1,2,3,4,5-pentaphosphate (D or L isomer was not designated) and subsequently removes phosphate from carbon 5, 4 or 1, and 3 to yield myoinositol-2-phosphate as the final product. The mechanism of dephosphorylation of phytate by mung bean phytase appears to be different from that of F2 phytase of wheat bran.

Most of the studies on plant phytases present a consensus view that the phytate is the natural substrate for phytase with myoinositol-2-phosphate as its ultimate dephosphorylation

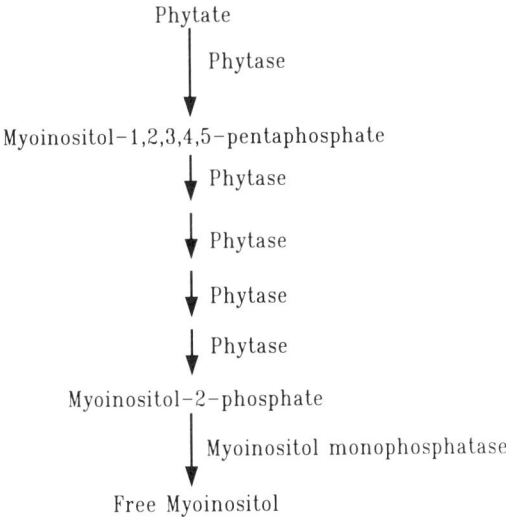

FIGURE 3. Dephosphorylation of phytate by phytase.[20,22]

product.[15,18] However, Biswas et al.[26] and De and Biswas[38] presented a separate proposal, wherein a portion of phytate is dephosphorylated differently during germination of mung bean seeds (Figure 2). They reported that a phytate-ADP-phosphotransferase specifically removes phosphate from carbon 2 of phytate to yield as an initial product myoinositol-1,3,4,5,6-pentaphosphate. Subsequently, a phytase like the F2 phytase of wheat bran acts on this pentaphosphate to produce myoinositol-1-phosphate as the final product. A myoinositol monophosphatase is needed for complete hydrolysis of myoinositol-1-phosphate or myoinositol-2-phosphate to generate free myoinositol and inorganic phosphate, both of which are metabolically in demand during germination of seeds. Yamagata et al.[47] reported that an acid phosphatase from aleurone particles of rice grain may play a role in the hydrolysis of myoinositol monophosphate, a final product of phytate dephosphorylation. Loewus and Loewus[48] isolated an alkaline Mg^{++}-dependent myoinositol phosphatase that hydrolyzes myoinositol-2-phosphate as well as 1D or 1L-myoinositol-1-phosphate. Probably, myoinositol monophosphatase provides the final step in phytate hydrolysis; however, careful examination of additional enzymes with phytase activity, especially with regard to their developmental occurrence, cellular localization, and substrate specificities, is needed to better describe the pathway for phytate degradation during germination.[15]

Biswas et al.[26] proposed that myoinositol-1-phosphate generated during dephosphorylation of phytate may be converted to glucose-6-phosphate and used for phytate biosynthesis. De and Biswas[38] isolated a novel enzyme system, myoinositol-1-phosphate dehydrogenase from germinating mung bean seeds. This enzyme catalyzes the conversion of myoinositol-1-phosphate to ribulose-5-phosphate, a pentose phosphate intermediate.[49] Based on this, a novel metabolic cycle may be presented involving glucose-6-phosphate and myoinositol phosphates during formation and germination of seeds[38] (Figure 2). During the germination of the seed the presence of myoinositol-1-phosphate dehydrogenase system provides a link between the metabolic pathway of the inositol phosphates and the pentose phosphate cycle. The pentose phosphate cycle is reported to be very active during the early period of germination.[38]

Strother[50] has proposed a homeostatic mechanism for utilization of endogenous phytate in germinating seeds, whereby germinating seeds maintain a steady state concentration of inorganic phosphate. Phytate degradation by phytase may be regulated by phytase biosynthesis,[51,53] phytase inhibition by inhibitors such as inorganic phosphate[32,52,54,59] and cycloheximide,[53,55] relative activity of phytase on partially dephosphorylated forms of phytate,[20] phytase activation,[43] and hormones such as gibberellic acid.[56,57] Scott and Loewus[15] reviewed the role of hormones on phytate degradation.

Phytate can also be dephosphorylated by heating in acid solutions at pH 3.0 to 4.0.[17] Dephosphorylation in neutral or alkaline solution appears to be very slow. Hydrolysis of phytate at acidic pH 4.0 results in a more complex mixture of lower phosphates than does enzymatic hydrolysis.[17]

REFERENCES

1. **Sobolev, A. and Rodionova, M. A.**, Phytin synthesis by aleurone grains in ripening sunflower seeds, *Soviet Plant Physiol.* (English transl.), 13, 958, 1966.
2. **Sobolev, A.**, On the state of phytin in the aleurone grains of mature and germinating seeds, *Soviet Plant Physiol.* (English transl.), 13, 177, 1966.
3. **Ogawa, M., Tanaka, K., and Kasai, Z.**, Phytic acid formation in dissected ripening rice grains, *Agric. Biol. Chem.*, 43, 2211, 1979.
4. **Ogawa, M., Tanaka, K., and Kasai, Z.**, Accumulation of phosphorus, magnesium, and potassium in developing rice grains: followed by electron microprobe x-ray analysis focusing on the aleurone layer, *Plant Cell Physiol.*, 20, 19, 1979.
5. **Tanaka, K., Yoshida, T., Asada, K., and Kasai, Z.**, Subcellular particles isolated from aleurone layer of rice seeds, *Arch. Biochem. Biophys.*, 155, 136, 1973.
6. **Tanaka, K., Yoshida, T., and Kasai, Z.**, Radioautographic demonstration of the accumulation site of phytic acid in rice and wheat grains, *Plant Cell Physiol.*, 15, 147, 1974.
7. **Tanaka, K., Ogawa, M., and Kasai, Z.**, The rice scutellum: studies by scanning electron microscopy and electron microprobe x-ray analysis, *Cereal Chem.*, 53, 643, 1976.
8. **Tanaka, K., Yoshida, T., and Kasai, Z.**, Phosphorylation of myoinositol by isolated aleurone particles of rice, *Agric. Biol. Chem.*, 40, 1319, 1976.
9. **Tanaka, K., Nishitomi, T., Ogawa, M., Yoshida, T., and Kasai, Z.**, Formation of adenosine triphosphate in isolated aleurone particles of rice grains, *Agric. Biol. Chem.*, 40, 1313, 1976.
10. **Lott, J. N. A. and Buttrose, M.**, Globoids in protein bodies of legume seed cotyledons, *Aust. J. Plant Physiol.*, 5, 89, 1978.
11. **Tanaka, K. and Kasai, Z.**, Phytic acid in rice grains, in *Antinutrients and Natural Toxicants in Foods*, Ory, R. L., Ed., Food and Nutrition Press, Westport, Connecticut, 1981, 239.
12. **Sasaki, K. and Loewus, F. A.**, Metabolism of myo-[2-^{3}H] inositol and scyllo-[R-^{3}H] inositol in ripening wheat kernels, *Plant Physiol.*, 66, 740, 1980.
13. **Lott, J. N. A.**, Accumulation of seed reserves of phosphorus and other minerals, in *Seed Physiology*, Vol. 1, Murray, D. R., Ed., Academic Press, New York, 1984, 139.
14. **Lott, J. N. A. and Ockenden, I.**, The fine structure of phytate-rich particles in plants, in *Phytic Acid: Chemistry and Applications*, Graf, E., Ed., Pilatus Press, Minneapolis, 1986, 43.
15. **Scott, J. J. and Loewus, F. A.**, Phytate metabolism in plants, in *Phytic Acid: Chemistry and Applications*, Graf, E., Ed., Pilatus Press, Minneapolis, 1986, 23.
16. **Asada, K., Tanaka, K., and Kasai, Z.**, Formation of phytic acid in cereal grains, *Ann. N.Y. Acad. Sci.*, 165, 801, 1969.
17. **Cosgrove, D. J.**, The chemistry and biochemistry of inositol polyphosphates, *Rev. Pure Appl. Chem.*, 16, 209, 1966.
18. **Cosgrove, D. J.**, *Inositol Phosphates: Their Chemistry, Biochemistry, and Physiology*, Elsevier, New York, 1980, 118.
19. **Loewus, F. A. and Loewus, M.W.**, Myo-inositol: biosynthesis and metabolism, in *Biochemistry of Plants*, Vol. 3, Preiss, J., Ed., Academic Press, New York, 1980, 43.
20. **Loewus, F. A. and Loewus, M. W.**, Myo-inositol: its biosynthesis and metabolism, *Annu. Rev. Plant. Physiol.*, 34, 137, 1983.
21. **Loewus, F. A. and Dickinson, D. B.**, Cyclitols, in *Encyclopedia of Plant Physiology*, Vol. 13A, Loewus, F. A. and Tanner, W., Eds., Springer-Verlag, New York, 1982, 193.
22. **Loewus, F. A.**, Phytate metabolism, with special reference to its myoinositol component, in *Recent Advances in Phytochemistry*, Vol. 17, Nozzolillo, C., Lea, P. J., and Loewus, F. A., Eds., Plenum Press, New York, 1983, 173.

23. **Asada, K., Tanaka, K., and Kasai, Z.,** Phosphorylation of myoinositol in ripening grains of rice and wheat. Incorporation of phosphate-^{32}P and myoinositol-^{3}H into myoinositol phosphates, *Plant Cell Physiol.,* 9, 185, 1968.
24. **Mandal, N. C. and Biswas, B. B.,** Metabolism of inositol phosphates. II. Biosynthesis of inositol polyphosphates in germinating seeds of *Phaseolus aureus, Indian J. Biochem.,* 7, 63, 1970.
25. **Chakrabarti, S. and Majumder, A. L.,** Phosphoinositol kinase from plant and avian sources, in *Cyclitols and Phosphoinositides,* Wells, W. W. and Eisenberg, F., Jr., Eds., Academic Press, New York, 1978, 69.
26. **Biswas, B. B., Biswas, S., Chakrabarti, S., and De, B. P.,** A novel metabolic cycle involving myoinositol phosphates during formation and germination of seeds, in *Cyclitols and Phosphoinositides,* Wells, W. W. and Eisenberg, F., Jr., Eds., Academic Press, New York, 1978, 57.
27. **Ahuja, J. N.,** Studies on the biosynthesis of phytic acid, University Microfilm 61-1634, Ann Arbor, Michigan, 1962.
28. **Asada, K. and Kasai, Z.,** Formation of myoinositol and phytin in ripening rice grains, *Plant Cell Physiol.,* 3, 397, 1962.
29. **Igaue, I., Shimizu, M., and Miyauchi, S.,** Formation of a series of myoinositol phosphates during growth of rice plant cells in suspension culture, *Plant Cell Physiol.,* 21, 351, 1980.
30. **Igaue, I., Miyauchi, S., and Saito, K.,** Formation of myoinositol phosphates in a rice cell suspension culture, in *Proc. 5th Int. Congr. Plant Tissues and Cell Culture,* Fujiwara, A., Ed., 1982, 265.
31. **Igaue, I. and Miyauchi, S.,** Occurrence of organic acids from myoinositol in cultured rice cells, *Agric. Biol. Chem.,* 46, 1413, 1982.
32. **Roberts, R. W. and Loewus, F. A.,** Inositol metabolism in plants. VI. Conversion of myoinositol to phytic acid in *Wolffiella floridana, Plant Physiol.,* 43, 1710, 1968.
33. **Inhulsen, D. and Niemeyer, R.,** Inositol phosphates from *Lemna minor* L., *Z. Pflanzenphysiol.,* 88, 103, 1978.
34. **Graf, E.,** Formation of (^{3}H ^{32}P) phytic acid in germinating wheat, *Anal. Biochem.* 131, 351, 1983.
35. **Majumder, A. N. L., Mandal, N. C. and Biswas, B. B.,** Phosphoinositol kinase from germinating mung bean seeds, *Phytochemistry,* 11, 503, 1972.
36. **Majumder, A. N. L. and Biswas, B. B.,** Further characterization of phosphoinositol kinase isolated from germinating mung bean seeds, *Phytochemistry,* 12, 315, 1973.
37. **Chakrabarti, S. and Biswas, B. B.,** Two forms of phosphoinositol kinase from germinating mung bean seeds, *Phytochemistry,* 20, 1815, 1981.
38. **De, B. P. and Biswas, B. B.,** Evidence for the existence of a novel enzyme system: myoinositol-1-phosphate dehydrogenase in *Phaseolus aureus, J. Biol. Chem.,* 254, 8717, 1979.
39. **Tanaka, K., Watanabe, K., Asada, K., and Kasai, Z.,** Occurrence of myoinositol monophosphate and its role in ripening rice grains, *Agric. Biol. Chem.,* 35, 314, 1971.
40. **Loewus, M. W., Sasaki, K., Leavitt, A. L., Munsell, L., Sherman, W. R., and Loewus, F. A.,** Enantiomeric form of myoinositol-1-phosphate produced by myoinositol-1-phosphate synthase and myoinositol kinase in higher plants, *Plant Physiol.,* 70, 1661, 1982.
41. **Biswas, S., Maiti, S. B., Chakrabarti, S., and Biswas, B. B.,** Purification and characterization of myoinositol hexaphosphate-adenosine diphosphate-phosphotransferase from *Phaseolus aureus, Arch. Biochem. Biophys.,* 185, 557, 1978.
42. **Tomlinson, R. V. and Ballou, C. E.,** Myoinositol polyphosphate intermediates in the dephosphorylation of phytic acid by phytase, *Biochemistry,* 1, 166, 1962.
43. **Lim, P. E. and Tate, M. E.,** The phytases. I. Lysolecethin-activated phytase from wheat bran, *Biochim. Biophys. Acta,* 250, 155, 1971.
44. **Lim, P. E. and Tate, M. E.,** The phytases. II. Properties of phytase fractions F_1 and F_2 from wheat bran and the myo-inositol phosphates produced by fraction F_2, *Biochim. Biophys. Acta,* 302, 316, 1973.
45. **Maiti, I. B., Majumder, A. N. L., and Biswas, B. B.,** Purification and mode of action of phytase from *Phaseolus aureus, Phytochemistry,* 13, 1047, 1974.
46. **Maiti, I. B. and Biswas, B. B.,** Further characterization of phytase from *Phaseolus aureus, Phytochemistry,* 18, 316, 1979.
47. **Yamagata, H., Tanaka, K., and Kasai, Z.,** Isoenzymes of acid phosphatase in aleurone particles of rice grains and their interconversions, *Agric. Biol. Chem.,* 43, 2059, 1979.
48. **Loewus, M. W. and Loewus, F. A.,** Myoinositol-1-phosphatase from the pollen of *Lilium longiflorum* Thunb., *Plant Physiol.,* 70, 765, 1982.
49. **Ghosh, B., De, B. P., and Biswas, B. B.,** Purification and properties of myoinositol-1-phosphate dehydrogenase from germinating mung bean seeds, *Arch. Biochem. Biophys.,* 228, 309, 1984.
50. **Strother, S.,** Homeostasis in germinating seeds, *Ann. Bot.,* 45, 217, 1980.
51. **Bianchetti, R. and Sartirana, M. L.,** The mechanism of the repression by inorganic phosphate of phytase synthesis in the germinating wheat embryo, *Biochem. Biophys. Acta,* 145, 485, 1967.
52. **Sartirana, M. L. and Bianchetti, R.,** The effect of phosphate on the development of phytase in the wheat embryo, *Physiol. Plant.,* 20, 1066, 1967.

53. **Mandal, N. C. and Biswas, B. B.,** Metabolism of inositol phosphates. I. Phytase synthesis during germination in cotyledons of mung beans, *Phaseolus aureus, Plant Physiol.,* 45, 4, 1979.
54. **Mandal, N. C. Burman, S., and Biswas, B. B.,** Isolation, purification, and characterization of phytase from germinating mung beans, *Phytochemistry* 11, 495, 1972.
55. **Kuvaeva, E. B. and Kretovich, V. L.,** Phytase of germinating pea seeds, *Soviet Plant Physiol.,* 25, 290, 1978.
56. **Eastwood, D. and Laidman, D. L.,** The mobilization of macronutrient elements in germinating wheat grain, *Phytochemistry,* 10, 1275, 1971.
57. **Katayama, N. and Suzuki, H.,** Possible effect of gibberellin on phytate degradation in germinating barley seeds, *Plant Cell Physiol.,* 21, 115, 1980.
58. **Phillippy, B. Q., White, K. D., Johnston, M. R., Tao, S. H., and Fox, M. R. S.,** Preparation of inositol phosphates from sodium phytate by enzymatic and nonenzymatic hydrolysis, *Anal. Biochem.,* 162, 115, 1987.
59. **Beal, L. and Mehta, T.,** Zinc and phytate distribution in peas: influence of heat treatment, germination, pH, substrate, and phosphorus on pea phytate and phytase, *J. Food Sci.,* 50, 96, 1985.
60. **Majerus, P. W., Connolly, T. M., Bansal, V. S., Inhorn, R. C., Ross, T. S., and Lips, D. L.,** Inositol phosphates: synthesis and degradation, *J. Biol. Chem.,* 263, 3051. 1988.
61. **Greenwood, J. S. and Bewley, J. D.,** Subcellular distribution of phytin in the endosperm of developing castor bean: a possibility for its synthesis in the cytoplasm prior to deposition within protein bodies, *Planta,* 160, 113, 1984.
62. **Organ, M. G., Greenwood, J. S., and Bewley, J. D.,** Phytin is synthesized in the cotyledons of germinated castor bean seeds in response to exogenously supplied phosphate, *Planta,* 174, 513, 1988.

Chapter 3

PHYTASE ENZYME: BIOSYNTHESIS AND CHARACTERIZATION

I. INTRODUCTION

Phytase is an enzyme with esterase activity which is capable of hydrolyzing myoinositol hexakisphosphate to inorganic orthophosphate and a series of lower phosphoric esters of myoinositol and, in some cases, to free myoinositol. The enzyme nomenclature accepted by the International Union of Pure and Applied Chemistry and the Union of Biochemistry,[1] recognizes two phytases: a 3-phytase which catalyzes the following reaction:

$$\text{myoinositol } 1,2,3,4,5,6 \text{ hexakisphosphate} + H_2 \rightarrow \text{myoinositol } 1,2,3,4,5 \text{ pentakisphosphate} + \text{orthophosphate}$$

and a 6-phytase which catalyzes the following reaction:

$$\text{myoinositol } 1,2,3,4,5,6 \text{ hexakisphosphate} + H_2O \rightarrow \text{myoinositol } 1,2,3,4,5 \text{ pentakisphosphate} + \text{orthophosphate}$$

The 3-phytase (EC 3.1.3.8) appears to be characteristic of microorganisms, while the 6-phytase (EC 3.1.3.26) is found in seeds of higher plants.[2] There have been extensive studies on phytases from a variety of sources: plants such as soybeans,[3,4] navy beans,[5] mung beans,[6-9] dwarf French beans,[10] faba beans,[11,12] California small white beans,[13,14] peanuts,[15] *Vigna* beans,[16] garden pea,[17] triticale,[18] wheat,[19-23] corn,[24] barley,[25] rice,[26-28] and sorghum,[29] as well as animals such as rats, chickens, calves, humans,[30] pigs,[31] and microbial sources.[32-35]

The first published report on the preparation of phytase was by Suzuki et al.[26] The presence of phytases in animal tissues was first reported for calf liver and blood[36] and the occurrence of phytase in rat intestine was first noted by Patwardhan.[32] Bitar and Rienhold[30] have shown that phytase activity is present in human intestine. Several other researchers later detected phytase activity in rat intestine[37-39] and chicken intestine.[40-42] It is, however, not known whether or not this is of any nutritional consequence *in vivo* with respect to utilization of phytate as a source of phosphorus. Davies and Flett[39] reported that the rat intestinal phytase and alkaline phosphatase were one and the same. Their conclusion was based on the following observations: (1) Both activities were similarly distributed in the small intestine, the greatest amount in the duodenum and lowest in the terminal ileum. The regional differences in activity were reflected by similar differences in the capacity to hydrolyze phytate *in vivo* by the ligated intestinal segments. (2) Both enzymes had tenfold greater activities in the brush border fraction of duodenal mucosa compared with the whole mucosal homogenates. (3) Brush border activities of both the enzymes required magnesium and zinc ions for maximal activity. (4) Zinc deficiency induced by feeding low (0.5 mg Zn/kg) dietary zinc caused similar decline in the activity of both the enzymes. To date, however, no studies have been reported to provide conclusive evidence that both these activities are the same.

Recent studies have shown that phytase is present in rat intestine[43] and small intestine of rat, rabbit, guinea pig, and hamster.[44] The former study found two phytase activities (pH optima 4.7 and 8.0) that were distinctly different than phosphatase activity in the rat intestine (although the enzymes were not purified). These complexities arise in part due to the difficulty in separating the phytase activity from the phosphatase activity.

II. PHYTASE SYNTHESIS AND LOCATION IN SEEDS

During seed germination, phytate is rapidly degraded and there is a steady rise in phytase activity.[5,6,11,17,28,45-49,64] It is not clear whether this increase in phytase activity is due to synthesis of new phytase or simple activation of existing phytase. In mung bean[6] and wheat,[21] phytase is synthesized during germination. In contrast, others [50,51] found an increase in phytase activity in wheat due to the activation of existing phytase during germination. The mechanism for this activation has not been established.

Kuvayeva and Kretovich[52] used Sephadex CM-50 chromatography to purify phytase from both the dry and germinated peas. Electrophoresis studies with 7.5% acrylamide gels (glycine-acetic acid buffer, pH 4.0) indicated the presence of two distinct proteins with phytase activity. The form with higher mobility appeared exclusively during germination. Detailed biosynthetic pathways for phytase synthesis have not been elucidated.

Studies on determining phytases in seeds have shown that in ungerminated wheat 34.1% of phytase is associated with the endosperm, 15.3% with scutellum, and 39.5% with aleurone tissue.[19] In barley, phytase has been found in the fine structures surrounding protein bodies in the aleurone layer;[53] in sorghum, with the spherosomes of the protein bodies in the aleurone layer;[29,54] and in rice, with aleurone particles[27] in the aleurone layer.[55] Similar data in legumes is lacking.

III. PHYTASE EXTRACTION AND PURIFICATION

Distilled water,[19,45,49,56-58] buffer solutions,[10,54] and salt solutions[3,5,11] have been used to extract phytase from plants. The crude extracts are then typically subjected to ammonium sulfate precipitation followed by dialysis or gel filtration to remove the salt, and then to ion exchange chromatography. Methanol[56] and acetone[19,47,56] precipitation have also been used to concentrate the phytase activity. The highest purification (1500 fold) of a plant phytase has been obtained from wheat bran.[20] Recently, Gibson and Ullah[65] purified two phytases from cotyledons of germinating soybeans to near homogeneity.

IV. MOLECULAR PROPERTIES OF PHYTASES

The molecular weight of only a few phytase preparations has been reported. Using gel filtration, Mandal et al.[8] estimated the molecular weight of 1.6×10^5 Daltons for mung bean phytase, and Yamamoto and co-workers[59] estimated the molecular weight of *Aspergillus terreus* phytase to be 2.2 to 2.3×10^5 Daltons. In the presence of guanidine hydrochloride the latter enzyme dissociated into six monomers, each with a molecular weight of 3.7×10^4 Daltons.[59] Lim and Tate[22] estimated a molecular weight of $47,000 \pm 2000$ Daltons for wheat bran phytases F_1 and F_2. Scott and Loewus[60] recently reported the presence of a calcium-activated phytase (an alkaline form) in the pollen of *Lilium longiflorum Thunb.* with an estimated molecular weight of 88,000 Daltons.

Activation energies for several phytases are summarized in Table 1. Since the hydrolysis of myoinositol hexaphosphate is a stepwise process furnishing lower esters (which are also substrates for the phytases), the nature of substrate cannot be accurately defined. The values in Table 1, therefore, should be evaluated with caution.

Since different phosphate groups can be the first site of hydrolysis, the reaction can be further complicated. For example, Maiti et al.[7] showed that the degradation of phytate by mung bean phytase starts with the dephosphorylation of position 6 phosphate followed by phosphate removal from positions 5 and 4, 1 and 3, or 1 and 4; position 2 phosphate (axial phosphate group) being stable. Lim and Tate,[22] on the other hand, have shown that F_2 fraction from wheat bran phytase preparation hydrolyzes from positions D-4, -2, or -5. This suggests that there is a need

TABLE 1
Energy of Activation for Phytase[a]

Phytase source	Substrate	Activation energy (cal/mole)	Ref.
Navy bean	Myoinositol hexaphosphate	11,500	5
California Small White bean	Myoinositol hexaphosphate	9,200	14
Germinated mung bean	Myoinositol hexaphosphate	8,500	8
Wheat bran	Myoinositol hexaphosphate	12,000	56
Wheat bran	Myoinositol hexa- to myoinositol monophosphate	11,000	20
Soybean	Myoinositol hexaphosphate	11,100	3

[a] Since hydrolysis of myoinositol hexaphosphate is a stepwise process, the nature of substrate is not well defined and consequently these figures may represent average value for the initial stages of the hydrolysis.

TABLE 2
Temperature and pH Optima, and Km Values for Cereal and Legume Phytases

Phytase source	Optimum temperature (°C)	Optimum pH	Michaelis-Menten constant K_m (M)	Ref.
Triticale	45	5.4	0.22×10^{-3}	18
Corn	50	5.6	0.99×10^{-3}	24
Wheat flour	55	5.15	0.33×10^{-3}	19
Wheat bran	—	5.0	0.57×10^{-3}	56
Rice aleurone particles	45	4.0—5.0	—	27
Navy bean	50	5.3	0.018×10^{-3}	5
California Small White bean	60	5.2	0.22×10^{-4}	13
Dwarf French bean	40	5.2	0.65×10^{-3}	10
Mung beans (germinating)	57	7.5	0.65×10^{-3}	6
Faba beans (germinating)	50	5.0	0.017×10^{-3}	12
Soybean	60	4.8	2.4×10^{-3}	3
Wheat bran fraction F_1	—	5.6	2.2×10^{-3}	22
Wheat bran fraction F_2	—	7.2	2.0×10^{-4}	22

to understand the mechanism(s) for dephosphorylation of phytate, and whether or not the mechanism is dependent on the phytase source.

The temperature optima, pH optima, and the Km values for plant phytases are listed in Table 2. With the exception of mung beans and wheat bran fraction F_2, all the other plant phytases have a pH optimum range of 4.0 to 5.6. The optimum temperature range is 45 to 60°C. The Km values in this table need to be evaluated with caution because after the initial hydrolysis of phytate, the system becomes multisubstrate; therefore, the Km values can be correct only when the true initial rates are determined. Phytases are relatively stable when compared to several other enzymes (especially the proteolytic ones). This is reflected in their rather high temperature optima. Above 60°C, however, enzyme activity is rapidly lost and complete inactivation usually takes place at temperatures equal to or greater than 90°C.[3,18,56] Phytase activity in dry wheat bran has been shown to be stable up to 5 hours at 90°C.[61]

Compounds that activate and inhibit phytases are summarized in Tables 3 and 4, respectively. Divalent metal ions such as Ca^{2+}, Mg^{2+}, Co^{2+}, and Fe^{2+} seem to activate phytases, while fluoride is an inhibitor of plant phytases. The ability of *p*-chloromercuribenzoate to inhibit dwarf French bean phytase[10] and the inhibition of soybean phytase by *N*-ethylmaleimide and iodoacetamide[3] suggest that the thiol group may be an important factor in the expression of phytase activity in certain sources.

Relative substrate specificities of some cereal and legume phytases are shown in Table 5. As

TABLE 3
Activators of Phytase

Phytase source	Activator	Ref.
Navy bean	Co^{2+}	5
Corn endosperm scutellar tissue	Ca^{2+}	24
Wheat meal	Mg^{2+}, Ca^{2+}, NaN_3	19
Wheat bran	Mg^{2+}, Ca^{2+}	56
Wheat bran fraction F_1	Lysolecithin	58
Soybean	Fe^{2+}	3

TABLE 4
Inhibitors of Phytase

Phytase source	Inhibitor	Ref.
Triticale	Fe^{2+}, Fe^{3+}, Cu^{2+}, *p*-chloromercuribenzoate, Ni^{2+}, Co^{2+}, Ag^+	18
Corn endosperm scutellar tissue	F^-	24
Wheat meal	F^-, CN^-d, Zn^{2+}, Mn^{2+}, yeast extract	19
Wheat bran	F^-, $^+$, Hg^+, Ag^+	2, 56
Dwarf French bean	F^-, *p*-chloromercuribenzoate	10
Soybean	Zn^2, Cu^{2+}, Hg^+, *N*-ethylmaleimide, iodoacetamide, L-cysteine, L-ascorbic acid, 2-mercaptoethanol, citrate, oxalate, EDTA, tartrate	3
Faba bean	Zn^{2+}, Cu^{2+}, Fe^{2+}, Mg^{2+}, Co^{2+}, Ca^{2+}	12

TABLE 5
Relative Substrate Specificity of Cereal and Legume Phytases[5,8,56,57]

	Phytase source				
Substrate	Navy bean	Mung bean	Rice ear I	Rice ear III	Wheat fraction VII
P_6-myoinositol[a]	100	100	100	100	100
P_5-myoinositol[a]	—	235	—	—	—
P_4-myoinositol[a]	—	267	—	—	—
P_3-myoinositol[a]	—	293	—	—	—
P_2-myoinositol[a]	—	—	—	—	—
P_1-myoinositol[a]	—	—	—	—	—
β-glycerophosphate	1,530	33	44	92	65
α-glycerophosphate	1,900	—	—	—	90
Fructose diphosphate	—	—	—	—	130
Glucose-1-phosphate	—	117	40	13	0
Glucose-6-phosphate	—	—	—	—	50
5'-adenylic acid	970	—	13	17	40
Phenylphosphate	6,170	—	—	—	250
Pyrophosphate	73,500	1,300	58	133	460
Adenosine triphosphate	—	—	—	—	350
Adenosine diphosphate	—	—	—	—	220
Adenosine 5'-monophosphate	—	—	—	—	40
Triphosphopyridine nucleotide	—	—	—	—	90

[a] P_6-, P_5-, P_4-, P_3-, P_2-, and P_1-myoinositol refer to myoinositol hexa-, penta-, tetra-, tri-, di-, and mono-phosphates, respectively.

TABLE 6
Hydrolysis of Myoinositol Hexaphosphate Isomers by Wheat Bran Phytase[63]

Substrate	Relative activity
Myoinositol hexaphosphate	100
Neoinositol hexaphosphate	95
D-Chiroinositol hexaphosphate	47
Scylloinositol hexaphosphate	17

can be seen from this table, none of these enzyme preparations shows an absolute specificity for phytate. This is the reason why phytases have been frequently described as nonspecific acid phosphomonoesterases.[2] However, one report has described the presence of an acid phosphomonoesterase in dormant barley seed (*Hordeum vulgare* var. Kenia) protein bodies[62] that is specific for phytate. Phytases also catalyze hydrolysis of phytate isomers (Table 6), but to a lesser extent.[63]

REFERENCES

1. **IUPAC-IUB,** Enzyme nomenclature, *Biochim. Biophys. Acta,* 429, 1, 1976.
2. **Cosgrove, D. J.,** *Inositol Phosphates: Their Chemistry, Biochemistry, and Physiology,* Elsevier, New York, 1980, 85.
3. **Sutardi and Buckle, K. A.,** The characteristics of soybean phytase, *J. Food Biochem.,* 10, 197, 1986.
4. **Sudarmadji, S. and Markakis, P.,** The phytate and phytase of soybean tempeh, *J. Sci. Food Agric.,* 28, 381, 1977.
5. **Lolas, G. M. and Markakis, P.,** The phytase of navy beans *(Phaseolus vulgaris), J. Food Sci.,* 42, 1094, 1977.
6. **Mandal, N. C. and Biswas, B. B.,** Metabolism of inositol phosphates. I. Phytase synthesis during germination in cotyledons of mung beans *(Phaseolus aureus), Plant Physiol.,* 45, 4, 1970.
7. **Maiti, I. B., Majumdar, A. L., and Biswas, B. B.,** Purification and mode of action of phytase from *Phaseolus vulgaris, Phytochemistry,* 13, 1047, 1974.
8. **Mandal, N. C., Burman, S., and Biswas, B. B.,** Isolation, purification and characterization of phytase from germinating mung beans, *Phytochemistry,* 11, 495, 1972.
9. **Maiti, I. B. and Biswas, B. B.,** Further characterization of phytase from *Phaseolus aureus, Phytochemistry,* 18, 316, 1979.
10. **Gibbins, L. N. and Norris, F. W.,** Phytase and acid phosphatase in the dwarf bean (*Phaseolus vulgaris*), *Biochem. J.,* 86, 67, 1963.
11. **Eskin, N. A. M. and Wiebe, S.,** Changes in phytase activity and phytate during germination of two fababean cultivars, *J. Food Sci.,* 48, 270, 1983.
12. **Eskin, N. A. M. and Johnson, S.,** Isolation and partial purification of phytase from *Vicia faba* minor, *Food Chem.,* 26, 149, 1987.
13. **Chang, R. H.,** Removal of Phytic Acid from Beans by Potentiation of *In Situ* Phytase, Ph.D. dissertation, University of California, Berkeley, 1975.
14. **Chang, R. H. and Schwimmer, S.,** Characterization of phytase of beans *(Phaseolus vulgaris), J. Food Biochem.,* 1, 45, 1977.
15. **Davis, R. C., Jr.,** The purification and properties of peanut phytase and the identification of the myo-inositol phosphates from partial dephosphorylation of myoinositol hexaphosphate by the enzyme, Ph.D. thesis, Texas A&M Univ., College Station, Texas, 1968.
16. **Sugiura, M. and Sunobe, Y.,** Phosphorus compounds and phytase in germinating bean *Vigna sesquipedalis, Bot. Mag. (Tokyo),* 75, 65, 1962.
17. **Guardiola, J. L. and Sutcliffe, J. F.,** Mobilization of phosphorus in cotyledons of young seedlings of the garden pea *(Pisum sativum* L.), *Ann. Bot.,* 35, 809, 1971.
18. **Singh, B. and Sedeh, H. G.,** Characteristics of phytase and its relationship to acid phosphatase and certain minerals in triticale, *Cereal Chem.,* 56, 267, 1979.
19. **Peers, F. G.,** The phytase of wheat, *Biochem. J.,* 53, 102, 1953.

20. Nagai, Y. and Funahashi, S., Phytase from wheat bran, *Agric. Biol. Chem.,* 27, 619, 1963.
21. Sartirana, M. L. and Bianchetti, R., The effects of phosphate on the development of phytase in the wheat embryo, *Physiol. Plant.,* 20, 1066, 1967.
22. Lim, P. E. and Tate, M. E., The phytases. II. Properties of phytase fractions F_1 and F_2 from wheat bran and the *myo*-inositol phosphates produced by fraction F_2, *Biochim. Biophys. Acta,* 302, 316, 1973.
23. Ranhotra, G. S. and Loewe, R. J., Effect of wheat phytase on dietary phytic acid, *J. Food Sci.,* 40, 940, 1975.
24. Chang, C. W., Study of phytase and fluoride effects in germinating corn seeds, *Cereal Chem.,* 44, 129, 1967.
25. Preece, I. A. and Gray, H. J., Studies on phytin. II. Preliminary study of some barley phosphatases, *J. Inst. Brewing,* 68, 66, 1962.
26. Suzuki, U., Yoshimura, K., and Takaishi, M., Ueber ein Enzym "Phytase" das Anhydro-oxy-methylen disphosphorasure spaltet, *Coll. Agric. Bull. Tokyo Imp. Univ.,* 7, 495, 1907.
27. Yoshida, T., Tanaka, K., and Kasai, Z., Phytase activity associated with isolated aleurone particles of rice grain, *Agric. Biol. Chem.,* 39, 289, 1975.
28. Mukherji, S., Dey, B., Paul, A. K., and Sircar, S. M., Changes in phosphorus fractions and phytase activity of rice seeds during germination, *Physiol. Plant.,* 25, 94, 1971.
29. Adams, C. A. and Novellie, L., Acid hydrolases and autolytic properties of protein bodies and spherosomes isolated from ungerminated seeds of *Sorghum bicolor.* (Linn.) Moench, *Plant Physiol.,* 55, 7, 1975.
30. Bitar, K. and Reinhold, J. G., Phytase and alkaline phosphatase activities in intestinal mucosal of rat, chicken, calf, and man, *Biochim. Biophys. Acta,* 268, 442, 1972.
31. Pointillart, A., Fourdin, A., and Fontaine, N., Importance of cereal phytase activity for phytate phosphorus utilization by growing pigs fed diets containing triticale or corn, *J. Nutr.,* 117, 907, 1987.
32. Patwardhan, V. N., The occurrence of phytin-splitting enzyme in the intestines of albino rats, *Biochem. J.,* 31, 560, 1937.
33. Howson, S. J. and Davis, R. P., Production of phytate hydrolyzing enzyme by some fungi, *Enzyme Microb. Technol.,* 5, 377, 1983.
34. Sutardi and Buckle, K. A., Phytic acid changes in soybeans fermented by traditional inoculum and six strains of *Rhizoprus oligosporus, J. Appl. Bacteriol.,* 38, 538, 1985.
35. Wang, H. L., Swain, E. W., and Hesseltine, C. W., Phytase of molds used in oriental food fermentation, *J. Food Sci.,* 45, 1262, 1980.
36. McCollum E. V. and Hart, E. B., On the occurrence of a phytin-splitting enzyme in animal tissue, *J. Biol. Chem.,* 4, 497, 1908.
37. Pileggi, V. J., Distribution of phytase in the rat, *Arch. Biochem. Biophys.,* 80, 1, 1959.
38. Maddaiah, V. T., Kurnick, A. A., Hulett, B. J., and Reid, B. L., Nature of intestinal phytase activity, *Proc. Soc. Exp. Biol. Med.,* 115, 1054, 1964.
39. Davies, N. T. and Flett, A. A., The similarity between alkaline phosphatase (EC 3.1.2.1) and phytase (EC 3.1.3.8) activities in rat intestine and their importance in phytate-induced zinc deficiency, *Br. J. Nutr.,* 39, 307, 1978.
40. Davies, M. I., Ritchey, G. M., and Motzok, I., Intestinal phytase and alkaline phosphatase of chicks: influence of dietary calcium, inorganic and phytate phosphorus and vitamin D-3, *Poultry Sci.,* 49, 1280, 1970.
41. Davies, M. I. and Motzok, I., Properties of chick intestinal phytase, *Poultry Sci.,* 51, 494, 1972.
42. McCuaig, L. W., Davies, M. I., and Motzok, I., Intestinal alkaline phosphatase and phytase of chicks: effect of dietary magnesium, calcium, phosphorus and thyroactive casein, *Poultry Sci.,* 51, 526, 1972.
43. Ramakrishnan, C. V. and Bhandari, S. D., Differential developmental pattern of acid and alkaline phytase and phosphatase activities in rat intestine, *Experientia,* 35, 994, 1979.
44. Cooper, J. R. and Gowing, H. S., Mammalian small intestinal phytase (EC 3.1.3.8), *Br. J. Nutr.,* 50, 673, 1983.
45. Chen, L. H. and Pan, S. H., Decrease of phytates during germination of pea seeds *(Pisum sativa), Nutr. Rep. Int.,* 46, 125, 1977.
46. Ergle, D. R. and Guinn, G., Phosphorus compounds of cotton embryos and their changes during germination, *Plant Physiol.,* 34, 476, 1959.
47. Mayer, A. M., The breakdown of phytin and phytase activity in germinating lettuce seeds, *Enzymologia,* 19, 1, 1958.
48. Walker, K. A., Changes in phytic acid and phytase during early development of *Phaseolus vulgaris* L., *Planta,* 116, 91, 1974.
49. Bartnik, M. and Szafranska, I., Changes in phytate content and phytase activity during germination of some cereals, *J. Cereal Sci.,* 5, 23, 1987.
50. Eastwood, D., Tavener, R. J. A., and Laidman, D. L., Induction of lipase and phytase activities in the aleurone tissues of germinating wheat grains, *Biochem. J.,* 113, 32, 1969.
51. Eastwood, D. and Laidman, D. L., The mobilization of macronutrient elements in the germinating wheat grain, *Phytochemistry,* 10, 1275, 1971.
52. Kuvayeva, E. B. and Kretovich, V. L., Phytase of germinating pea seeds, *Soviet Plant Physiol.* (English transl.), 25, 290, 1978.

53. **Tronier, B., Ory, R. L., and Henningsen, K. W.,** Characterization of the fine structure and proteins from barley protein bodies, *Phytochemistry,* 10, 1207, 1971.
54. **Mayer, F. C., Campbell, R. E., Smith, A. K., and McKinney, L. L.,** Soybean phosphatase: purification and properties, *Arch. Biochem. Biophys.,* 94, 302, 1961.
55. **Palmiano, E. P. and Juliano, B. O.,** Changes in the activity of some hydrolases, peroxidase, and catalase in the rice seed during germination, *Plant Physiol.,* 52, 274, 1973.
56. **Nagai, Y. and Funahashi, S.,** Phytase (myo-inositol hexaphosphate phosphohydrolase) from wheat bran. I. Purification and substrate specificity, *Agric. Biol. Chem.,* 26, 794, 1962.
57. **Ikawa, T., Nisizawa, K., and Miwa, T.,** Specificities of several acid phosphatases from plant sources, *Nature,* 203, 939, 1964.
58. **Lim, P. E. and Tate, M. E.,** The phytases. I. Lysolecithin-activated phytase from wheat bran, *Biochim. Biophys. Acta,* 250, 155, 1971.
59. **Yamamoto, S., Minoda, Y., and Yamada, K.,** Chemical and physiochemical properties of phytase from *Aspergillus terreus, Agric. Biol. Chem.,* 36, 2097, 1972.
60. **Scott, J. J. and Loewus, F. A.,** A calcium-activated phytase from pollen of *Lilium longiflorum, Plant Physiol.,* 82, 33, 1986.
61. **McCance, R. A. and Widdowson, E. M.,** Activity of the phytase in different cereals and its resistance of dry heat, *Nature,* 153, 650, 1944.
62. **Ory, R. L. and Henningsen, K. W.,** Enzymes associated with protein bodies isolated from ungerminated barley seeds, *Plant Physiol.,* 44, 1488, 1969.
63. **Cosgrove, D. J.,** Synthesis of the hexaphosphates of myo-, scyllo-, and neo-, and D-inositol, *J. Sci. Food Agric.,* 17, 550, 1966.
64. **Roberts, R. M., Deshusses, J., and Loewus, F. A.,** Inositol metabolism in plants. V. Conversion of myoinositol to uronic acid and pentose units of acidic polysaccharides in root-tips of *Zea mays, Plant Physiol.,* 43, 979, 1968.
65. **Gibson, D. M. and Ullah, A. H. J.,** Purification and characterization of phytase from cotyledons of germinating soybean seeds, *Arch. Biochem. Biophys.,* 260, 503, 1988.

Chapter 4

PHYSIOLOGICAL FUNCTIONS AND USES OF PHYTATES

I. INTRODUCTION

Phytate is synthesized and deposited during seed development in discrete regions called globoids (also called aleurone particles).[1,2] It contains more than 80% of the total phosphorus in many edible legumes and cereal grains. Further, phytate accounts for a major portion of the stored reserves of phosphate and myoinositol in legumes and cereals[3-5] and is utilized as a source of phosphorus and myoinositol during germination of seeds, supports seedling growth, and supplies certain biosynthetic needs in the growing tissues. The young seedling utilizes the myoinositol as a substrate for the myoinositol oxidation pathway and ultimately for cell wall polysaccharides formation.[6] The presence of phytate in seeds and/or grains gives additional benefits to preservation of foods and protection against some diseases in humans. The physiological functions, benefits, and possible medical, industrial, and food applications of phytate are discussed in this chapter.

II. PHYSIOLOGICAL FUNCTIONS

Five physiological roles have been suggested for phytate in seeds and grains. These include (1) as a phosphorus store,[7,8] (2) as an energy store,[9-11] (3) as a source of cations,[12] (4) initiation of dormancy,[13] and (5) as a source of myoinositol (a cell wall polysaccharide precursor).[5] Cosgrove[4] extensively reviewed some of these physiological roles of phytate in seeds. Phytate may also act as a phosphagen during germination of seeds.[8] In addition to these, phytate may serve several additional unknown functions in seeds.

Besides having a role in the physiological function of the seed or grain, phytate may have an effect on the production of aflatoxin in infected seeds and grains. Gupta and Venkatasubramanian[14] reported that phytate in soybeans decreases aflatoxin production by *Aspergillus parasiticus*. They hypothesized that the zinc necessary for aflatoxin production is bound to phytate and thus makes it unavailable to the mold. However, Ehrlich and Ciegler[15-17] examined this hypothesis and concluded that phytate in soybeans is not responsible for its resistance to aflatoxin formation. They[16,17] suggested two possible explanations for inhibition of aflatoxin formation in soybeans. They are (1) presence of a low molecular weight substance in soybeans that is soluble in polar solvents and (2) pH. At higher pH values, phytate binds with zinc to form a complex that may be unavailable to the mold for aflatoxin production, while at lower pH values, no matter how much phytate is present, zinc will still be available.

Recently Graf et al.[18,19] suggested that phytate may serve as a natural antioxidant in seeds during dormancy. The antioxidant property of phytate is based on the assumption that the phytate effectively blocks iron-driven hydroxyl radical formation and suppresses lipid peroxidation.[19] Furthermore, the presence of phytate may slow oxidation of whole plant tissues and prevent the development of "putrid" odors possibly by depriving spoilage microorganisms of iron. This is especially advantageous in the case of fruits and vegetables wherein the browning and "putrefaction" can be prevented by the presence of high concentrations of phytate.[19]

A relationship also exists between cookability and phytate content of legumes.[20,21] With a higher phytate content there is a decreased cooking time in legumes; phytate removes calcium and magnesium ions and prevents formation of calcium and magnesium cross-linkages between pectate molecules of the middle lamellar tissues.

Graf and Eaton[22] suggested that diets rich in phytate suppress colonic carcinogenesis and other inflammatory bowel diseases by inhibiting intracolonic hydroxyl radical generation via

TABLE 1
Commercial Phytic Acid Specifications[27]

Product form	40—50% aqueous solution
Appearance	Colorless to light-brown syrup
Phosphorus content[a] (%)	26.0—29.0
Inorganic phosphorus (%)	<1.0
Salts (%)	<0.04
Sulfate (%)	<0.07
Heavy metals (%)	<0.004

[a] Dry weight basis.

From Sands, S. H., Biskobing, S. J., and Olson, R. M., in *Phytic Acid: Chemistry and Applications*, Graf, E., Ed., Pilatus Press, Minneapolis, 1986, 119. With permission.

the chelation of reactive iron. Sharma[23] reviewed the epidemiological evidence on the effects of dietary phytate in the prevention of diseases such as coronary heart disease, renal calculi, and colon cancer. The incidence of these diseases is high in the people of developed countries where high-fat, low-cereal diets are consumed. However, in developing countries where these diseases are less prevalent, low-fat, high-cereal diets containing coarse grains that are rich in phytate are consumed. Sharma[23] concluded that more research is required on the possible role of other dietary and environmental factors in the causation of these diseases before phytate can be recommended for controlling these diseases. Recently, Nielsen et al.[24] found that phytate decreases carcinogenesis in rats by decreasing colonic epithelial cell proliferation. Other researchers[25,26] have also reported that phytate reduces colon tumor incidence in rats. However, any use of phytate as a therapeutic agent needs to be carefully considered because of the adverse effects associated with its large intakes.[40-42]

III. COMMERCIAL MANUFACTURE AND USES

Sands et al.[27] reviewed the commercial availability, production, and potential applications of phytic acid and its salts. Corn, rice and wheat brans, cottonseed meal, etc., provide major sources for commercial production of phytic acid.[28] Corn steepwater produced during wet-milling of corn is a practical source for manufacture of phytic acid in the U.S. Rice bran is the primary starting material for phytic acid production in the Far East.[27] Currently, there are no major manufacturers of phytic acid in the U.S. All of the phytic acid and its salts sold in the U.S. are either imported or toll-manufactured. Phytic acid and its salts are mainly distributed through chemical supply companies. Phytic acid is sold as a 40.0 to 50.0% aqueous solution and is available in several different grades.[27] A typical phytic acid product specification is presented in Table 1. Phytic acid manufactured from corn steepwater contains about 26.5% phosphorus on a dry weight basis.[27]

Sands et al.[27] and Graf[28,29] summarized the food, industrial, and medical applications of phytic acid and suggested several additional uses for phytic acid and its salts. The food applications of phytic acid and its salts have been primarily realized as a result of the chelating and/or antioxidant properties of this compound. In the U.S., phytic acid and its salts are not permitted for use as an additive or preservative in foods. However, outside the U.S., phytic acid is extensively used as a food additive[28,30,31] to preserve a variety of foods. Addition of phytic acid to canned seafoods, fruits, and vegetables prevents product discoloration and improves the quality and shelf-life of cheese, noodles, miso, soy sauce, alcoholic drinks, fruit juices, and bread.[28,30,31] Phytic acid combined with lecithin has been patented as an antioxidant in Japan.[32] Phytic acid has been reported to inhibit oxidation of ascorbic acid, stabilize sorbic acid and prevent peroxidation and hydrolysis of fats and oils.[28] Phytic acid can be added to soybean oil and other lipid-containing foods to prevent both autoxidation and hydrolysis and to extend product quality and shelf-life. Phytate can be used to remove metals, especially iron, from liquid food products such as wines and other beverages.[28] Saio et al.[33] reported that phytic acid coagulates soybean protein in the presence of calcium and magnesium and improves the texture and flavor of tofu. Another major potential use of phytic acid is in the promotion of fermentations, especially in the pH range 6.0 to 8.0.[27] This could be useful in a wide variety of commercial fermentations including antibiotics,[34] yeast production,[35] lactic acid, and enzymes.[27]

The iron chelating properties of phytic acid provide for a wide variety of industrial applications.[27-29] These include the use of phytic acid as: (1) an anti-corrosion (or rustproofing) agent, especially useful in antifreeze, cooling water, or other closed systems for corrosion protection; (2) a replacement for cyanide or ammonium phosphate in etching solutions for offset printing in Japan; (3) a water additive for preventing scale formation on the walls of commercial reactors, boilers, and cooling towers; (4) an iron-stabilizing agent in the paper industry to maximize the efficiency of hydrogen peroxide bleaching of pulp; (5) a cation scavenger and catalyst for the polymerization of olefins; (6) a solvent (50 to 95% aqueous phytic acid) for the preparation of polyamide and polyurethane fibers of good tensile strength; (7) an antistatic agent in polymerization of ethylene to produce polymers with good transparency; and (8) an iron-chelator in preparation of lithographic plates.

A number of medical applications have been derived from phytic acid based on its interaction properties. However, some of these applications require confirmation through further biochemical and clinical experimentation.[28,29] Phytic acid binds to hydroxyapatite (a chief structural element of vertebrate bone and tooth made up of calcium phosphate) to form a uniform monomolecular surface layer that inhibits both dissolution and growth of hydroxyapatite crystals, resulting in inhibition of several physiological processes: (1) bone resorption, (2) bone turnover, (3) enamel dissolution, (4) cariogenesis, (5) plaque formation, and (6) growth of renal calculi. Based on the cariostatic properties, several commercial oral care products containing phytic acid have been formulated and patented, including dentifrices, mouth rinses, cleaning agents for dentures, and an adhesive film for removing nicotine tar from teeth.[27] Kaufman[36] reported that an oral mouth rinse containing phytic acid reduced plaque formation in humans. Phytic acid can be used in place of phosphoric acid for making strong dental silicate and zinc oxide cements.[37,38] Phytic acid has also been used as a metal chelator in prevention of scale formation and renal calculi, inhibition of platelet aggregation, and prevention and treatment of oxy-radical-induced diseases.[29] Wise[39] reported on the use of calcium phytate as a natural protective agent for lead poisoning mice. Phytate may provide some protection against toxic metal absorption in humans. Additional food, industrial, and medical applications of phytic acid and its salts have been discussed in detail in earlier reviews.[28-30]

REFERENCES

1. **Lott, J. N. A. and Buttrose, M. K.,** Globoids in protein bodies of legume seed cotyledons, *Aust. J. Plant Physiol.,* 5, 89, 1978.
2. **Lott, J. N. A. and Ockenden, I.,** The fine structure of phytate-rich particles in plants, in *Phytic Acid: Chemistry and Applications,* Graf, E., Ed., Pilatus Press, Minneapolis, 1986, 43.
3. **Cosgrove, D. J.,** The chemistry and biochemistry of inositol phosphates, *Rev. Pure Appl. Chem.,* 16, 209, 1966.
4. **Cosgrove, D. J.,** *Inositol Phosphates: Their Chemistry, Biochemistry, and Physiology,* Elsevier, New York, 1980, 139.
5. **Scott, J. J. and Loewus, F. A.,** Phytate metabolism in plants, in *Phytic Acid: Chemistry and Applications,* Graf, E., Ed., Pilatus Press, Minneapolis, 1986, 23.
6. **Loewus, F. A.,** Phytate metabolism with special reference to its myo-inositol component, in *Recent Advances in Phytochemistry,* Vol. 17, Nozzolillo, C., Lea, P. J., and Loewus, F. A., Eds., Plenum Press, New York, 1983, 173.

7. **Hall, J. R. and Hodges, T. K.**, Phosphorus metabolism of germinating oat seeds, *Plant Physiol.*, 41, 1459, 1966.
8. **Asada, K., Tanaka, K., and Kasai, Z.**, Formation of phytic acid in cereal grains, *Ann. N.Y. Acad. Sci.*, 165, 801, 1969.
9. **Biswas, S. and Biswas, B. B.**, Enzymatic synthesis of guanosine triphosphate, *Biochim. Biophys. Acta*, 108, 710, 1965.
10. **Biswas, S., Chakrabarti, S., and Biswas, B. B.**, Purification and characterization of myo-inositol hexaphosphate-adenosine diphosphate phosphotransferase from *Phaseolus aureus*, *Arch. Biochem. Biophys.*, 185, 557, 1978.
11. **Atkinson, M. R., and Morton, R. K.**, Free energy and the biosynthesis of phosphates, in *Comparative Biochemistry*, Vol. 2, Florkin, M. and Mason, H. S., Eds., Academic Press, New York, 1960, 1.
12. **Williams, S. G.**, The role of phytic acid in the wheat grain, *Plant Physiol.*, 45, 376, 1970.
13. **Sobolev, A. M. and Rodionova, M. A.**, Phytin synthesis by aleurone grain in ripening sunflower seeds, *Soviet Plant Physiol.* (English transl.), 13, 958, 1966.
14. **Gupta, S. K. and Venktasubramanian, T. A.**, Production of aflatoxin in soybeans, *Appl. Microbiol.*, 29, 834, 1975.
15. **Ehrlich, K. and Ciegler, A.**, Effect of phytate on aflatoxin formation by *Aspergillus parasiticus* and *Aspergillus flavus* in synthetic media, *Mycopathologia*, 87, 99, 1984.
16. **Ehrlich, K. and Ciegler, A.**, Effect of phytate on aflatoxin formation by *Aspergillus parasiticus* grown on different grains, *Mycopathologia*, 92, 3, 1985.
17. **Ehrlich, K. and Ciegler, A.**, Phytic acid and aflatoxin metabolism, in *Phytic Acid: Chemistry and Applications*, Graf, E., Ed., Pilatus Press, Minneapolis, 1986, 321.
18. **Graf, E., Mahoney, J. R., Bryant, R. G., and Eaton, J. W.**, Iron-catalyzed hydroxyl radical formation: stringent requirement for free iron coordination site, *J. Biol. Chem.*, 259, 3620, 1984.
19. **Graf, E., Empson, K. L. and Eaton, J. W.**, Phytic acid: a natural antioxidant, *J. Biol. Chem.*, 262, 11647, 1987.
20. **Kon, S. and Sanshuck, D. W.**, Phytate content and its effect on cooking quality of beans, *J. Food Process. Preser.*, 5, 169, 1981.
21. **Hincks, M. J. and Stanley, D. W.**, Multiple mechanisms of bean hardening, *J. Food Technol.*, 21, 731, 1986.
22. **Graf, E. and Eaton, J. W.**, Dietary suppression of colonic cancer: fiber or phytate, *Cancer*, 56, 717, 1985.
23. **Sharma, R. D.**, Phytate and the epidemiology of heart disease, renal calculi and colon cancer, in *Phytic Acid: Chemistry and Applications*, Graf, E., Ed., Pilatus Press, Minneapolis, 1986, 161.
24. **Nielsen, B. K., Thompson, L. U., and Bird, R. P.**, Effect of phytic acid in colonic epithelial cell proliferation, *Cancer Lett.*, 37, 317, 1987.
25. **Elsayed, A., Ullah, A., and Shamsuddin, A.**, Post-initiation dietary supplementation with corn-derived inositol hexaphosphate (IP) inhibits large intestinal carcinogenesis in Fisher 344 rats, *Fed. Proc.*, 46, 585, 1987.
26. **Elsayed, A., Chakravarthy, A., and Shamsuddin, A.**, Inositol hexaphosphate from corn decreased the frequency of colorectal cancer in azoxymethane-treated rats, *Lab. Invest.*, 56, 21A, 1987.
27. **Sands, S. H., Biskobing, S. J., and Olson, R. M.**, Commercial aspects of phytic acid: an overview, in *Phytic Acid: Chemistry and Applications*, Graf, E., Ed., Pilatus Press, Minneapolis, 1986, 119.
28. **Graf, E.**, Applications of phytic acid, *J. Am. Oil Chem. Soc.*, 60, 1861, 1983.
29. **Graf, E.**, Chemistry and applications of phytic acid: an overview, in *Phytic Acid: Chemistry and Applications*, Graf, E., Ed., Pilatus Press, Minneapolis, 1986, 1.
30. **Hayashi, K.**, Application of phytic acid to foods, *Kanazume Jiho (Jpn.)*, 58, 1006, 1979.
31. **Kato, A.**, Use of phytic acid in the production of noodles, *New Food Ind. (Jpn.)*, 22, 17, 1980.
32. **Tsukagoshi, K., Noda, S., and Tsuki, K.**, Preventing autoxidation of fats and oils, Japanese Patent, 724,401, 1972.
33. **Saio, K., Koyama, E., Yamazaki, S., and Watanabe, T.**, Protein-calcium-phytic acid relationship in soybean, *Agric. Biol. Chem.*, 53, 36, 1969.
34. **Yamaguchi, K., Saio, K., Suzuki, T., and Nakayama, K.**, Stimulation of fermentative production of aminoglycoside antibiotics by addition of phytins and phytic acid, Japanese Patent, 7,638,490, 1976.
35. **Kawano, T., Arima, S., Kojima, H., Mizumoto, S., and Morinaga, K.**, Cultivation of yeasts for use in feed, Japanese Patent 7,305,989, 1973.
36. **Kaufman, H. W.**, Interaction of inositol phosphates with mineralized tissues, in *Phytic acid: Chemistry and Applications*, Graf, E., Ed., Pilatus Press, Minneapolis, 1986, 303.
37. **Prosser, H. J., Brant, P. J., Scott, R. P., and Wilson, A. D.**, The cement-forming properties of phytic acid, *J. Dent. Res.*, 62, 598, 1983.
38. **Prosser, H. J. and Wilson, A. D.**, The cement-forming properties of phytic acid, in *Phytic Acid: Chemistry and Applications*, Graf, E., Ed., Pilatus Press, Minneapolis, 1986, 291.
39. **Wise, A.**, Protective action of calcium phytate against acute lead toxicity in mice, *Bull. Environ. Contam. Toxicol.*, 27, 630, 1981.
40. **Thompson, L. U.**, Antinutrients and blood glucose, *Food Technol.*, 42, 123, 1988.
41. **Harland, B. F. and Oberleas, D.**, Phytate in foods, *World Rev. Nutr. Diet.*, 52, 235, 1987.
42. **Cheryan, M.**, Phytic acid interactions in food systems, *CRC Crit. Rev. Food Sci. Nutr.*, 13, 297, 1980.

Chapter 5

METHODS FOR ANALYSIS OF PHYTATE

I. INTRODUCTION

Mature legumes, cereals, and their milled fractions and/or by-products contain the highest concentrations of phytate. Accurate methods are needed for the determination of phytate in legumes and cereals. There are no known specific reagents that identify phytate, nor does it have a characteristic absorption spectrum. Most of the analytical methods for phytate are based on extraction or isolation, purification, and subsequent measurement of the stoichiometric ratios with ferric ion or of the phosphate component of phytate.[3] Qualitative and quantitative methods for analysis of phytate have been extensively reviewed.[1-3]

II. QUALITATIVE METHODS

Qualitative methods are mainly used for separation and subsequent identification of inositol phosphate esters. For details of individual qualitative methods and their applications the readers are referred to reviews by Oberleas[1] and Oberleas and Harland.[3]

A. PAPER CHROMATOGRAPHY

Paper chromatography is one of the most popular methods for separating and identifying various inositol phosphate esters from a mixture. Effective separation of inositol phosphate esters by paper chromatography depends on (1) selection of chromatographic papers free of heavy metal ions or the removal of heavy metals from the paper since heavy metal impurities will bind to phosphate esters and cause trailings, (2) selection of proper eluant mixture, and (3) detection of inositol phosphate esters after separation.[3] Several methods involving washing of chromatographic papers with various chemical solutions have been developed for the elimination of heavy metal impurities from chromatographic papers.[4-9] For instance, Anderson[4,5] washed the chromatographic paper with $2N$ HCl solution containing 0.5% ethylenediaminetetraacetate (EDTA) followed by ion-free water for removal of heavy metal impurities. Johnson and Tate[9] used a wash solution containing 0.1 M oxalic acid for 17 hours followed by a 1.5M ammonia rinse.

Hanes and Isherwood[6] surveyed 60 different eluant mixtures and found only 7 useful for separating inositol phosphate esters. Desjobert and Petek[11] developed a satisfactory eluant mixture based on n-propanol:concentrated ammonia:water (5:4:1). This eluant mixture has been modified and extensively used for separation of inositol phosphate esters.[9,10]

Inositol phosphate esters separated by paper chromatograms are usually detected using one of the many modifications of the Hanes and Isherwood[6] method. The most convenient modification was described by Harrap.[12] In this method, the dried paper is dipped in a freshly prepared detection mixture of perchloric acid (60%, 5 ml), ammonium molybdate solution (20% w/v, 5 ml), HCl (1.0N, 10 ml), and acetone (80 ml). The paper is first dried for 5 min in air, and then exposed to UV light, dipped in 2.5% alpha-benzoinoxime in methanol, and then air dried. Under acid conditions, alpha-benzoinoxime reacts with free molybdenum to provide a white background that persists for several weeks. Phosphate esters appear as blue areas, becoming more intense after a period of exposure in diffuse daylight. Inositol phosphate esters may be recovered from paper chromatograms by using the techniques of Wade and Morgan[13] and Sobolev.[14]

B. PAPER ELECTROPHORESIS

Wade and Morgan[13] were the first to apply paper electrophoresis for separation of inositol phosphate esters. They treated papers with N formic acid to remove interfering substances that would otherwise retard movement of polyphosphates. In this system, the paper was impregnated with an aqueous solution of butyric acid (9.2% v/v) and NaOH (0.1%), and exposed to a potential of 400 V applied across their length for 4.5 hours at room temperature. Wade and Morgan[13] also demonstrated the feasibility of combining electrophoresis with chromatography for providing two-dimensional separations.

Arnold[15] applied a slightly modified electrophoresis technique to the separation of inositol phosphates (especially separation of hexa- and pentaphosphates). The modifications he made included supporting the paper between two plate glass sheets, utilizing a $0.2N$ acetate buffer (pH 3.65), and a potential of 220 V across a 50-cm paper for 16 h. Seiffert and Agranoff[16] and Roberts and Loewus[17] modified this technique to achieve improved separations of inositol phosphate esters from a mixture.

Tate[18] combined the moving paper electrophoresis method with the oxalate buffer (pH 1.50) system[16] and succeeded in the complete separation of inositol pentaphosphates from a mixture. The components on the paper following electrophoresis are detected by the methods described for the paper chromatography.

C. THIN-LAYER CHROMATOGRAPHY

Thin-layer chromatography is less suitable for the separation of inositol phosphate esters because of the polar nature of the phosphate esters. Angyal and Russell[19] overcame these disadvantages by methylating inositol phosphate esters with diazomethane in anhydrous methanol and were able to separate two pentaphosphate esters on silica gel (Silica Gel H) using anhydrous methanol as the solvent. The separated phosphate esters were detected by exposing the plates to iodine vapors.

D. ION EXCHANGE CHROMATOGRAPHY

Smith and Clark[20] separated a mixture of inositol phosphates by ion exchange chromatography using a weak-base exchange resin (De-Acidite) and stepwise elution with 0.1 to $0.8N$ HCl. Further, they separated the hydrolysis products of inositol phosphates into a number of components and identified them by phosphorus:inositol ratios: anion exchange resins Dowex 1 (Cl^-),[3,21,22] Dowex 2 (Cl^-),[8,23,24] and Dowex AG1-X8 (Cl^-, 200 to 400 mesh)[7,25] and linear gradient eluting systems consisting of either HCl or water. NaCl ($1M$) has been used to separate the intermediate inositol phosphates from phytate following partial hydrolysis by acid or phytase.

III. QUANTITATIVE METHODS

The most common methods used for the quantitative determination of phytate are derived from the procedure developed by Heubner and Stadler.[26] This method is based on the principle that phytate forms an insoluble complex with ferric ion in dilute acid solution and presumably is the only phosphate compound with that property. Inorganic phosphate is not precipitated under these conditions and organic phosphates other than phytate also remain in solution. In the original procedure, phytate containing extracts were titrated with standardized ferric chloride solution using ammonium thiocyanate as an internal indicator to detect the first excess of ferric ion in solution. A pink color produced by ferric thiocyanate that persisted for 5 min served as the endpoint. However, the major disadvantage of this method is the difficulty of detecting a consistent endpoint in the presence of the precipitating white ferric phytate and/or reducing substances and other organic particles contained in the extract.[3] There have been numerous modifications of this original method.[27-33]

Amperometric, complexometric titration, phytase, and inositol methods developed for quantitative analysis of phytate have never received popular acceptance.[34-36] Harland and

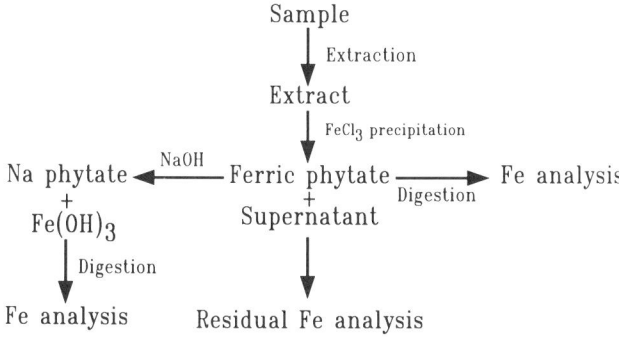

FIGURE 1. Indirect methods for phytate analysis.

Oberleas[37] published a method for quantifying the phytate content in foods by incorporating step gradient elution and anion exchange chromatography. This method has been widely accepted[2,3,38-40] and served as a basis for the development of several other methods.[41-44] High-performance liquid chromatography (HPLC) and Fourier transform-nuclear magnetic resonance (FT-NMR) procedures for the quantitation of phytate have been developed.[45-50]

Extraction is the first step in any procedure for the quantitative measurement of phytate in legumes, cereals, and their products. The most common phytate extractants are hydrochloric acid (0.6, 1.2, and 2.4%)[26,28,32,37,39-41,49,51-53] and trichloroacetic acid (3 and 5%).[45,47,48,51,52,54-57] Earley and DeTurk[32] and Thompson and Erdman[51,52] added 10% sodium sulfate to 1.2% hydrochloric acid extract for producing a firmer ferric phytate precipitate and improving its retention during subsequent washing steps. Use of trichloroacetic acid as an extractant has some advantages over hydrochloric acid because trichloroacetic acid effectively denatures and coagulates the proteins in the extract, thus preventing possible interference with phytate at later stages of phytate measurement. Nitric acid ($0.5N$) has been used for extraction of phytate from dried peas.[33] Recently, Camire and Clydesdale[46] reported that 3% sulfuric acid is more effective in extracting phytate from wheat bran than trichloroacetic acid (3%) or hydrochloric acid (1%).

According to Oberleas and Harland,[3] fat content influences the extractability of phytate from food products. The fat content of the material should be kept low (<5%) or reduced before extraction of phytate with dilute aqueous acid. Quantitative methods for phytate are presented under three categories: precipitation methods, chromatographic methods, and NMR and other methods.

A. PRECIPITATION METHODS

The precipitation methods have evolved from the original titration method of Heubner and Stadler.[26] These methods can be divided into indirect and direct methods (Figures 1 and 2). In the indirect method, a known standard quantity of ferric chloride in dilute acid is added to the extract to precipitate phytate and either the portion of standardized ferric iron in ferric phytate or the portion of the unprecipitated ferric iron in the solution is measured by a standard colorimetric method. A stoichiometric relationship between iron and phosphorus (ratio of 4:6 Fe:P) is used to calculate phytate content. In the direct method, insoluble stable ferric phytate precipitate is removed and the phosphorus content of the precipitate is determined after wet ashing or hydrolysis; alternatively, inositol content of the precipitate is determined.

1. Indirect Methods

Young[31] first introduced the indirect method for phytate determination. This method is based on the precipitation of ferric phytate with a slight excess of a standardized solution of ferric chloride in dilute hydrochloric acid. The excess unprecipitated ferric iron is measured colorimet-

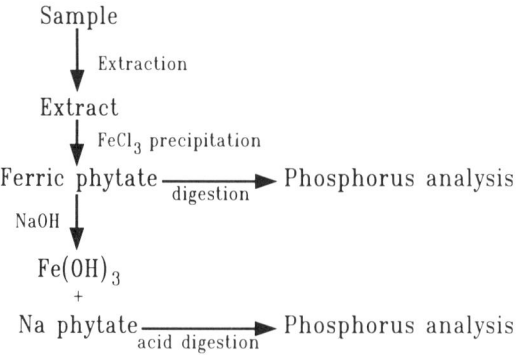

FIGURE 2. Direct methods for phytate analysis.

rically as ferric thiocyanate after removing the ferric phytate precipitate (Figure 1). On the other hand, deLange et al.[55] collected and washed the ferric phytate precipitate and measured the iron in the dissolved precipitate colorimetrically. In both cases, phytate content is calculated by using a conversion factor derived from a Fe:P ratio of 4:6. Young[31] used an experimentally determined factor of 1.06, whereas the theoretical factor is 1.20. Subsequently, Earley[58] found that under carefully controlled conditions, the theoretical value of the conversion factor is applicable. Many of the researchers[31,51,54,58-63] have experimentally obtained Fe:P ratios of 3.5 to 4.6:6 in determining the phytate content in several food products, seeds, and grains. However, a ratio of 4:6 (Fe:P) is generally used for calculating phytate content and it favors the formation of 4Fe-phytate.[64,65] In order to achieve this ratio during precipitation, a 3.3 to 3.6-fold excess of ferric ion must be present in the reaction mixture.[3,58]

The indirect methods appear to be more convenient and rapid than direct methods. However, when the phytate level in the food under examination is low, the indirect method is subject to large errors. Samotus and Schwimmer[59] reported that the results for the phytate content of freshly prepared acid extracts of plant material as determined by the indirect method was erroneously high. This high result was possibly due to the presence of substances (ascorbic acid and chlorogenic acid) which reduced Fe^{3+} to Fe^{2+}. Addition of 30% hydrogen peroxide prior to precipitation of phytate with ferric chloride was reported to overcome this difficulty.

The presence of excess iron in the extract (either by too much addition of ferric chloride or due to naturally occurring high iron content in the seed or grain itself) shifts the equilibrium resulting in the formation of soluble ferric phytate, thus leading to the underestimation of phytate content.[66] According to the findings of Anderson,[66] recovery of phytate in barley was only 80% when the soluble iron to phytate phosphorus ratio was about 0.33:1.0.

The indirect method has also been used in the microdetermination of phytate[61] in the solution or bound to hydroxyapatite in concentrations up to 10^{-5} M. In spite of some undesirable features, the indirect method has been widely used for measuring phytate in cereals and legumes.[54,60,67,68] Wheeler and Ferrel[54] extracted phytate in ground cereals with 3% trichloroacetic acid and heated the extract with a ferric chloride solution to precipitate phytate. Ferric phytate was converted to ferric hydroxide (insoluble precipitate) and sodium phytate (soluble) by adding sodium hydroxide and boiling. The iron content of ferric hydroxide later was determined colorimetrically.

Recently, Haug and Lantzch[68] simplified the indirect method for the rapid determination of phytate in cereals and cereal products. In this method, an aliquot of extract is heated with a standardized acidic iron-III solution and the iron content in the supernatant determined. The decrease of iron in the supernatant is proportional to the phytate content (3 to 30 µg/ml) of the cereal. However, this method needs to be calibrated with standard phytic acid solutions for each set of analysis.

Thompson and Erdman[52] reported that the phytate methods that rely on the analysis of iron from the ferric phytate precipitate are not recommended. Their conclusion was based primarily on the variations observed in the Fe:P molar ratio of ferric phytate precipitated from soybean extracts.

2. Direct Methods

McCance and Widdowson[30] introduced the direct method for quantitation of phytate in foodstuffs. This method is based on the precipitation of ferric phytate with excess ferric chloride solution in dilute hydrochloric acid. Ferric phytate is converted to soluble sodium phytate and ferric hydroxide precipitate using sodium hydroxide. The phosphorus content of the soluble sodium phytate is then determined colorimetrically after wet-ashing with concentrated sulfuric/perchloric acids (Figure 2). The time requirement for the acid digestion in this method is, however, undesirable. Several researchers[32,69-71,80] modified the direct method to improve the precision and to reduce the time required for analysis. For instance, Earley and DeTurk[32] washed the ferric phytate precipitate with diluted acid: sulfate solution, moistened with 50% magnesium nitrate, dried, then dry-ashed at 1000°C for 1 hour. The phosphorus content in the dry-ashed sample is determined colorimetrically (Figure 2). Heggen and Reith[69] suggested hydrolysis of soluble sodium phytate and direct estimation of the resulting inositol by a periodate oxidation method. Ellis et al.[80] reported that the presence of a high concentration of inorganic phosphate in the extract may lead to contamination of the ferric phytate precipitate and result in high phytate values. The errors due to high inorganic phosphate can be eliminated by incubating the ferric phytate in $0.5M$ HCl and heating.

Several other researchers have developed methods for quantitation of phytate without using wet-acid hydrolysis. Latta and Eskin[41] developed a simple method for the rapid direct colorimetric determination of phytate based on the reaction between ferric chloride and sulfosalicylic acid. They eluted an acidic extract through an anion-exchange resin to remove inorganic phosphorus and other interfering compounds. The phytate content in the eluted extract was directly measured by using modified Wade reagent (0.03% $FeCl_3 \cdot 6H_2O$ and 0.3% sulfosalicylic acid in distilled water). Recently, Mohamed et al.[72] developed another direct spectrophotometric method to quantitate phytate without acid digestion. This method is based on the precipitation of phytate as ferric phytate followed by conversion to soluble sodium phytate. A modified chromogenic reagent is then mixed with soluble sodium phytate, heated for 30 min at 95°C and the blue color complex formed measured at 830 nm. This method may be useful in monitoring chromatographic separation of inositol phosphates.

B. CHROMATOGRAPHIC METHODS

1. Ion Exchange Chromatographic Methods

Although this method has been used extensively for qualitative studies and separation of inositol phosphates,[7,20,21,24,25] Caldwell and Black[73] modified the procedure for the quantitation of inositol hexaphosphates in extracts of soils and manures. Extracts were dissolved in $0.85N$ hydrochloric acid (HCl) and phytates were separated with a De-Acidite anion exchange resin (60 to 80 mesh) using step gradients of 0.85, 1.4, and $3.0N$ HCl. They discarded the first $0.85N$ HCl eluate and the inositol hexaphosphate in the next $1.4N$ and $3.0N$ HCl eluates was determined by analysis of the phosphate content after wet-ashing. Other researchers[7,20] reported that 0.8-$1.0N$ HCl or salt (sodium chloride) would elute inositol hexaphosphates from extracts under the above conditions. Harland and Oberleas[37] quantified phytate in textured vegetable proteins using anion-exchange resin (AG1-X8 chloride form, 200 to 400 mesh) and step gradients of $0.05M$ and $0.7M$ sodium chloride (NaCl). In this method, the HCl extracts were diluted and passed through an anion-exchange resin column. The major contaminant, inorganic phosphate, remained on the column and was washed with $0.05M$ NaCl. Phytate was eluted with $0.7M$ NaCl. The final eluate containing phytate was wet-ashed and analyzed for phosphorus colorimetri-

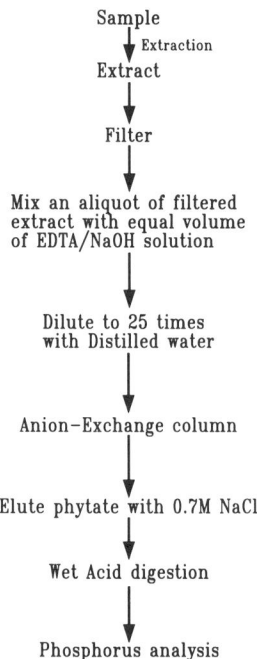

FIGURE 3. Flow chart for phytate analysis by ion exchange chromatography.

cally. Small quantities of phytate could be retained on the resin and determined by this method. Cosgrove[75] employed a slightly modified ion exchange method for the quantitative estimation of phytate from plant products. A trichloroacetic acid (TCA) extract of sample was diluted and applied to an anion-exchange resin (Dowex 1-X2, Cl⁻ form, 200 to 400 mesh) column and all of the phosphoric esters other than phytate were eluted with $0.1M$ TCA. The phytate from the column was then eluted with $1.0M$ HCl and analyzed for phosphorus content following wet-acid digestion. There have been several other modifications of the ion exchange resin method in order to improve sensitivity, separation, and reproducibility at low concentrations of phytate.[38-40] The original anion-exchange resin (AG1-X8, 200 to 400 mesh, Cl⁻ form)[37] was not satisfactory for elution of phytate from extracts of certain samples. Ellis and Morris[40] evaluated several lots of AG1-X8 (200 to 400 mesh), AG1-X8 (100 to 200 mesh), and AG1-X4 (100 to 200 mesh) anion-exchange resins for recovery of phytate. A wide variation in recovery of phytate was observed with different lots of AG1-X8 (100 to 200 mesh) anion-exchange resin. They found that the AG1-X4 (100 to 200 mesh) anion-exchange resin was most appropriate for both accuracy and rapidity in phytate analysis.

Other modifications made recently were in the acid extracts of the sample. Ellis and Morris[39] indicated that the addition of a small amount of EDTA to the sample extract and adjustment of the pH of the extract to 6.0 would improve the recovery of phytate. Addition of EDTA to the sample extract eliminates cations which otherwise would interfere with phytate elution.

Considering the above modifications, Harland and Oberleas[74] developed an anion-exchange method (Figure 3) for determination of phytate in foods. The modified method was collaboratively studied and adopted as "official new method" by the Association of Official Analytical Chemists (AOAC) in 1985. Briefly, the new method is described as follows:

Phytate from the dried sample is extracted with 2.4% HCl for 3 h and the mixture is vacuum-filtered through filter paper. An aliquot of acid extract is mixed with an equal volume of EDTA/NaOH ($0.11M$ disodium EDTA, $0.75M$ NaOH) solution. The extract is diluted to 25 times with distilled water for decreasing total anion concentration ($\sim 0.1M$) and quantitatively transferred

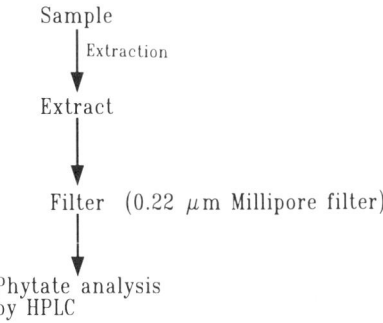

FIGURE 4. Phytate analysis by HPLC.

to a column containing anion-exchange resin (AG1-X4, 100 to 200 mesh, Cl⁻ form). Inorganic phosphate is eluted with 0.1M NaCl and later the phytate is eluted with 0.7M NaCl and collected. The phytate fraction is digested with a mixture of concentrated nitric and sulfuric acids and the phytate content is determined colorimetrically. The amount of phytate in the sample is calculated as hexaphosphate equivalent.

2. High Performance Liquid Chromatographic Methods

Tangendjaja et al.[45,76] first used high performance liquid chromatography (HPLC) with a differential refractive index detector for the analysis of phytate (Figure 4) in rice bran. The column employed was Bondapak C_{18} (30 cm × 4 mm). Without additional preparation they injected the filtered (0.22 µm Millipore® filter) sample extract of 3% TCA (trichloroacetic acid) into the column and eluted phytate with 5 mM sodium acetate as the mobile phase at a flow rate of 0.5 to 2.0 ml/min. The phytate was eluted from the column in 2 to 4 min depending on the flow rate before TCA. Since then, several investigators[42-44,46,47,77] have used HPLC equipped with C_{18} column with modifications in sample extract preparations and composition of mobile phase for better separation and improving the sensitivity of phytate at low concentrations. Camire and Clydesdale[46] used a 5 cm precolumn containing silica to protect the analytical C_{18} column from deterioration due to strong basic solutions. Most researchers filtered sample extracts through a 0.22 µm or 0.45 µm Millipore® filter prior to injection of the sample for HPLC separation.

Graf and Dintzis[42,77] modified the Tangendjaja et al.[45] method by adding a prepurification step to the sample extract preparation. They extracted samples with 0.5M HCl and neutralized the HCl extracts with 0.125N NaOH. The neutralized HCl extracts were passed through an anion-exchange resin (AG1-X8, 200 to 400 mesh, Cl⁻ form) column to remove inorganic phosphate and other impurities (mono- and divalent cations and proteins). They[77] eluted inorganic phosphate and other impurities from the column with small aliquots of 0.1M NaCl. Phytate was then eluted from the column with 2.0N HCl, collected, and evaporated to dryness under vacuum. The dried sample residue was re-solubilized in 5 mM sodium acetate prior to HPLC injection. A 5 mM sodium acetate was used as a mobile phase. The phytate elutes immediately after the void volume with a retention time of 1.4 to 1.6 min. This procedure eliminates interferences from solvent components and is sensitive to as little as 2 µg and linear up to 10 mg of phytate.

Instead of sample extract prepurification to remove inorganic phosphate and other impurities, Knuckles et al.[47] used a 25 mM potassium phosphate monobasic (KH_2PO_4) (pH 6.0) as the mobile phase. These investigators extracted samples with 3% TCA. The sample extract was diluted and adjusted to pH 6.0 by adding 0.5M KH_2PO_4 (pH 6.0). The diluted sample extract was filtered through a 0.2 µm membrane filter and an aliquot of the filtrate was injected onto the HPLC column. A 25 mM KH_2PO_4 (pH 6.0) was used as the mobile phase. Phytate elutes with the solvent front at a retention time of 1.1 to 1.2 min. In this method, phytate was linear up to 7 µg. Further, Knuckles et al.[47] evaluated various buffers including formate (pH 2.6 and 3.0), acetate (pH 3.0 and 4.0), and KH_2PO_4 (pH 6.0 and 6.5) as mobile phase at concentrations from 25 to 100 mM for the effect of pH on phytate analysis.

In most HPLC methods, phytate elutes with the solvent front. To eliminate errors due to co-elution of phytate with the solvent front, Lee and Abendroth[43] further modified the HPLC procedure. They essentially followed the procedure of Graf and Dintzis,[42] including prepurification of the sample extract by ion-exchange chromatography. The freeze-dried sample extracts were re-solubilized in 10 ml of mobile phase (2.5, 5.0 or 10.0 mM tetrabutylammoniumhydroxide [TBA-OH] as an ion-pair) adjusted to pH 7.2 with formic acid. Increasing the concentrations of the ion-pair (i.e., TBA-OH as the mobile phase) caused shorter retention times. The void volume which contains sodium and other ions eluted in 1.3 min, whereas the phytate eluate had a retention time of 4.75 to 11.30 min, depending on the concentration of mobile phase. The shorter retention time for phytate represents a higher concentration of the ion-pair in the mobile phase. The detection limit was reported to be 2 ng and linearity was up to 40 µg of phytate by refractive index in this method.

Camire and Clydesdale[46] found considerable masking of the phytate peak by TCA when attempting to duplicate the Tangendjaja et al.[45] method with a different C_{18} column (25 cm × 4.6 cm ID containing spherisorb ODS C_{18}, 10µ packing). To resolve the masking problem, several preparatory steps were added and a new quantitative HPLC method was developed. In this method, sample extracts of 3% sulfuric acid (H_2SO_4) were boiled with ferric chloride to aid in precipitation of ferric phytate. Subsequently, ferric phytate was washed and converted to soluble sodium phytate and ferric hydroxide precipitate with the addition of sodium hydroxide and boiling. The soluble sodium phytate was made to a known volume and an aliquot of it was then injected into the HPLC column. A 5 mM sodium acetate solution was used as a mobile phase at a flow rate of 0.5 ml/min. The phytate with solvent front was eluted between 3.0 to 3.5 min. A detection limit of 5 µg and linearity up to 100 µg of sodium phytate were obtained in this method. A mixture of inositol tri-, tetra-, penta-, and hexaphosphate in foods and intestinal contents could be separated and quantitatively determined by HPLC using a reverse-phase C_{18} column.[44]

The HPLC methods presented above seem to have the sensitivity and reproducibility to measure low concentrations of phytate in foods. In the future, HPLC may become the method of choice for routine analysis of phytate in foods and foodstuffs. However, additional research needs to be carried out on HPLC methods for phytate including extractants, preparation techniques, different columns, mobile phases, and flow rates in order to have a widely accepted method for routine analysis.[3,44]

Phillippy and Johnston[53] adapted a low pressure ion-exchange flow system with the post-column reactor method of Fitchett and Woodruff[78] for analysis of phytate. This method also detects inositol tri-, tetra-, and pentaphosphates in addition to hexaphosphate in processed foods and food products.[53,81,82] In this system, an aliquot of filtered sample extract is directly injected into the column. Phytate is eluted with 0.11M HNO_3 (continuously purged with a slow stream of helium) as the mobile phase. The eluted phytate is mixed with 0.1% ferric nitrate in 2% perchloric acid in a post-column reactor and the optical density of the ferric phytate complex was determined at 290 nm wavelength. They[53] reported that complete elution of phytate occurs in 7 min. However, Oberleas and Harland[3] indicated that many substances, including nitrate ion, chloride ion, and many pigments (extracted by aqueous acids) also have the absorbance at or near 290 nm wavelength. They suggested the use of a 282 nm wavelength may be more appropriate for improving sensitivity of phytate from interfering substances.[3]

Cilliers and Van Niekerk[57] described another method for liquid chromatographic determination of phytate by postcolumn colorimetric detection. In this procedure, a filtered sample extract of 3% TCA was directly injected into the column. Separation of phytate was accomplished on an anion-exchange column, IC-PAK-A (5 cm × 0.46 cm ID) with 45 mM $NaNO_3$ containing 18

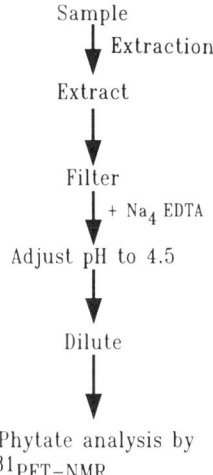

FIGURE 5. Phytate analysis using phosphorus[31] Fourier transform nuclear magnetic resonance spectrometry ([31]PFT-NMR).

mM HNO$_3$ as the mobile phase. The eluted phytate was detected at 500 nm by means of an in-line post-column reaction with a Wade reagent (0.015% FeCl$_3$ containing 0.15% sulfosalicylic acid pumped at 0.72 ml/min). This method is reported to be sensitive, reproducible, and specific, and requires only a simple sample preparation. The interferences from other substances may not be found in this method because of the specific reaction of the post-column Wade reagent with phosphates.

C. NUCLEAR MAGNETIC RESONANCE AND OTHER METHODS

O'Neill et al.[48] developed the phosphorus[31] Fourier transform nuclear magnetic resonance ([31]PFT-NMR) spectrometric technique for the direct quantitative estimation of phytate in foods. [31]PFT-NMR is reported to be specific for inositol hexaphosphates and discriminates lower inositol phosphates and inorganic phosphates. Food samples were extracted with 3% TCA containing 0.1% sodium cyclotetrametaphosphate tetrahydrate. The pH of the sample extract was adjusted to 4.5 ± 0.5 with the addition of tetrasodium EDTA, and filtered (0.45 μm Millipore filter). The viscosity of the pH adjusted sample extract was increased with sucrose for final NMR analysis. Mazzola et al.[49] improved the convenience and accuracy of the [31]PFT-NMR method (Figure 5) and eliminated interferences from paramagnetic ions. Excess quantities of EDTA were added to the pH adjusted sample extracts prior to NMR analysis to mask interferences from paramagnetic ions. EDTA preferentially binds paramagnetic ions in the aqueous acid extract and thereby effectively separates them from phytate. Further, Mazzola et al.[49] eliminated use of sucrose in sample extracts for increasing viscosity. This modified method claimed to be sensitive and specific for phytate analysis. The hydrolysis products of phytate such as lower inositol phosphates can also be measured by [31]PFT-NMR.[48,50]

Kikunaga et al.[79] quantitatively measured the phytate content of cereal grains by isotachophoresis (Figure 6). They followed essentially the direct precipitation method including conversion of ferric phytate to soluble sodium phytate. The isotachophoretic analyzer was used to measure the concentration of phytic acid in the soluble sodium phytate fraction. This method has a linearity ≥ 25 nM.

Titration and amperometric methods have also been developed for determination of phytate. Oberleas and Harland[3] presented an extensive review of these and other methods.

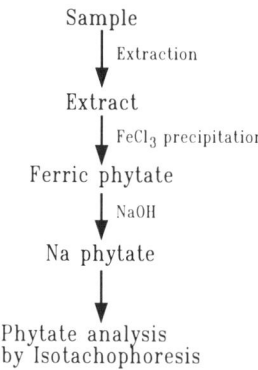

FIGURE 6. Phytate analysis using isotachophoresis.

REFERENCES

1. **Oberleas, D.,** The determination of phytate and inositol phosphates, in *Methods of Biochemical Analysis,* Vol. 20, Glick, D., Ed., John Wiley & Sons, Inc., New York, NY, 1971, 37
2. **Cosgrove, D. J.,** *Inositol Phosphates: Their Chemistry, Biochemistry, and Physiology,* Elsevier, Amsterdam, Netherlands, 1980, 12.
3. **Oberleas, D. and Harland, B. F.,** Analytical methods for phytate, in *Phytic Acid: Chemistry and Applications,* Graf, E., Ed., Pilatus Press, Minneapolis, MN, 1986, 77.
4. **Anderson, G.,** Paper chromatography of inositol phosphates, *Nature,* 175, 863, 1955.
5. **Anderson, G.,** Identification and estimation of soil inositol phosphates, *J. Sci. Food Agric.,* 7, 437, 1956.
6. **Hanes, C. S. and Isherwood, F. A.,** Separation of phosphoric esters on the filter paper chromatogram, *Nature,* 164, 1107, 1949.
7. **Cosgrove, D. J.,** The isolation of myoinositol pentaphosphates from hydrolysates of phytic acid, *Biochem. J.,* 89, 172, 1963.
8. **Kerr, S. E. and Kfoury, G. A.,** Chromatographic separation and identification of myo-inositol phosphates, *Arch. Biochem. Biophys.,* 96, 347, 1962.
9. **Johnson, L. F. and Tate, M. E.,** Structure of phytic acid, *Can. J. Chem.,* 47, 63, 1969.
10. **Pizer, F. L. and Ballou, C. E.,** Studies on myo-inositol phosphates of natural origin, *J. Amer. Chem. Soc.,* 81, 915, 1959.
11. **Desjobert, A. and Petek, F.,** Chromatographic sur papier des esters phosphoriques de l'inositol: application a l'etude de la degradation hydrolytique de l'inositol hexaphosphate, *Soc. Chim. Biol. Bull.,* 38, 871, 1956.
12. **Harrap, F. E. G.,** The detection of phosphate esters on paper chromatograms, *Analyst,* 85, 452, 1960.
13. **Wade, H. E. and Morgan, D. M.,** Fractionation of phosphates by paper ionophoresis and chromatography, *Biochem. J.,* 60, 264, 1955.
14. **Sobolev, A. M.,** Paper chromatography of inositol phosphates, *Fiziol. Rast.,* 9, 649, 1962.
15. **Arnold, P. W.,** Paper ionophoresis of inositol phosphates, with a note on the acid hydrolyzates of phytic acid, *Biochem. Biophys. Acta,* 19, 552, 1956.
16. **Seiffert, U. B. and Agranoff, B. W.,** Isolation and separation of inositol phosphates from hydrolyzates of rat tissues, *Biochim. Biophys. Acta,* 98, 574, 1956.
17. **Roberts, R. M. and Loewus, F.,** Inositol metabolism in plants. VI. Conversion of myo-inositol to phytic acid in *Wolffiella floridana, Plant Physiol.,* 43, 1710, 1968.
18. **Tate, M. E.,** Separation of myo-inositol pentaphosphates by moving paper electrophoresis (MPE), *Anal. Biochem.,* 23, 141, 1968.
19. **Angyal, S. J. and Russell, A. F.,** Cyclitols. XXVIII. Methyl esters of inositol phosphates and the structure of phytic acid, *Aust. J. Chem.,* 22, 383, 1969.
20. **Smith, D. H. and Clark, F. E.,** Chromatographic separations of inositol phosphorus compounds, *Proc. Soil Sci. Soc. Amer.,* 16, 170, 1952.
21. **Nagai, Y. and Funahashi, S.,** Phytase (myo-inositol hexaphosphate phosphohydrolase) for wheat bran. Part I. Purification and substrate specificity, *Agric. Biol. Chem.,* 26, 794, 1962.

22. **Tomlinson, R. V. and Ballou, C. E.,** Myo-inositol polyphosphate intermediates in the dephosphorylation of phytic acid by phytase, *Biochemistry,* 1, 166, 1962.
23. **Schormuller, J. and Bressau, G.,** Phosphate and organic phosphorus compounds in Foods. VIII. Preparative production of inosite phosphoric acid esters, *Z. Lebensm. Untersuch. U-Forsch.,* 113, 484, 1960.
24. **Mori, H.,** Phytin. II. Determination of phytin phosphorus, *Eiyo To Shokuryo,* 19, 235, 1966.
25. **Cosgrove, D. J.,** Ion-exchange chromatography of inositol polyphosphates, *Ann. N. Y. Acad. Sci.,* 165, 677, 1969.
26. **Heubner, W. and Stadler, H.,** Uber eine Titration — Methode zur Bestimmung des Phytins, *Biochem. Z.,* 64, 422, 1914.
27. **Rather, J. B.,** Determination of phytin phosphorus in plant products, *J. Amer. Chem. Soc.,* 39, 2506, 1917.
28. **Averill, H. P. and King, C. G.,** The phytin content of foodstuffs, *J. Amer. Chem. Soc.,* 48, 724, 1926.
29. **Harris, R. S. and Mosher, L. M.,** Estimation of phytin phosphorus, *Ind. Eng. Chem. (Anal. Ed.),* 6, 320, 1934.
30. **McCance, R. A. and Widdowson, E. M.,** Phytin in human nutrition, *Biochem. J.,* 29, 2694, 1935.
31. **Young, L.,** The determination of phytic acid, *Biochem. J.,* 30, 252, 1936.
32. **Earley, E. B. and DeTurk, E. E.,** Time and rate of synthesis of phytin in corn grain during the reproductive period, *J. Amer. Soc. Agron.,* 36, 803, 1944.
33. **Holt, R.,** Studies on dried peas: I. The determination of phytate phosphorus, *J. Sci. Food Agric.,* 6, 136, 1955.
34. **Babakhodzhaeva, S. A., Rizaev, M. U., and Gorokhouskaya, A. S.,** Determination of phytin in rice flour, *Pharm. Chem. J. (USSR),* 4, 42, 1968.
35. **Eagle, H., Agranoff, B. W., and Snell, E. E.,** The biosynthesis of meso-inositol by cultured mammalian cells and parabiotic growth of inositol-dependent and inositol-independent strains, *J. Biol. Chem.,* 235, 1891, 1960.
36. **Garcia-Villanova, R., Garcia-Villanova, R. J., and deLope, C. R.,** Determination of phytic acid by complexometric titration of excess iron (III), *The Analyst,* 107, 1503, 1982.
37. **Harland, B. F. and Oberleas, D.,** A modified method for phytate analysis using an ion-exchange procedure: application to textured vegetable proteins, *Cereal Chem.,* 54, 827, 1977.
38. **Ellis, R. and Morris, E. R.,** Comparison of ion-exchange and iron precipitation methods for analysis of phytate, *Cereal Chem.,* 59, 232, 1982.
39. **Ellis, R. and Morris, E. R.,** Improved ion-exchange phytate method, *Cereal Chem.,* 60, 121, 1983.
40. **Ellis, R. and Morris, E. R.,** Appropriate resin selection of rapid phytate analysis by ion-exchange chromatography, *Cereal Chem.,* 63, 58, 1986.
41. **Latta, M. and Eskin, N. A. M.,** A simple and rapid colorimetric method for phytate determination, *J. Agric. Food Chem.,* 28, 1313, 1980.
42. **Graf, E. and Dintzis, F. R.,** High-performance liquid chromatographic method for the determination of phytate, *Anal. Biochem.,* 119, 413, 1982.
43. **Lee, K. and Abendroth, J. A.,** High-performance liquid chromatographic determination of phytic acid in foods, *J. Food Sci.,* 48, 1344, 1983.
44. **Sandberg, A. S. and Ahderinne, R.,** HPLC method for determination of inositol tri-, tetra-, penta-, and hexaphosphates in foods and intestinal contents, *J. Food Sci.,* 51, 547, 1986.
45. **Tangendjaja, B., Buckle, K. A., and Wootton, M.,** Analysis of phytic acid by high-performance liquid chromatography, *J. Chromatography,* 197, 274, 1980.
46. **Camire, A. L. and Clydesdale, F. M.,** Analysis of phytic acid in foods by HPLC, *J. Food Sci.,* 47, 575, 1982.
47. **Kunckles, B. E., Kuzmicky, D. D. and Betschart, A. A.,** HPLC analysis of phytic acid in selected foods and biological samples, *J. Food Sci.,* 47, 257, 1982.
48. **O'Neill, I. K., Sargent, M., and Trimble, M. L.,** Determination of phytate in foods by phosphorus[31] Fourier transform nuclear magnetic resonance spectrometry, *Anal. Chem.,* 52, 1288, 1980.
49. **Mazzola, E. P., Phillippy, B. Q., Harland, B. F., Miller, T. H., Potemra, J. M., and Katsimpiris, E. W.,** Phosphorus[31] nuclear magnetic resonance spectroscopic determination of phytate in foods, *J. Agric. Food Chem.,* 34, 60, 1986.
50. **Frolich, W., Drakenberg, T., and Asp, N. G.,** Enzymatic degradation of phytate (myo-inositol hexaphosphate) in whole grain flour suspension and dough: a comparison between ^{31}PNMR spectroscopy and a ferric ion method, *J. Cereal Sci.,* 4, 325, 1986.
51. **Thompson, D. B. and Erdman, J. W.,** Phytic acid determination in soybeans, *J. Food Sci.,* 47, 513, 1982.
52. **Thompson, D. B. and Erdman, J. W.,** Structural model for ferric phytate: implications for phytic acid analysis, *Cereal Chem.,* 59, 525, 1982.
53. **Phillippy, B. Q. and Johnston, M. R.,** Determination of phytic acid in foods by ion chromatography with post-column derivatization, *J. Food Sci.,* 50, 541, 1985.
54. **Wheeler, E. L. and Ferrel, R. E.,** A method for phytic acid determination in wheat and wheat fractions, *Cereal Chem.,* 48, 312, 1971.
55. **deLange, D. J., Joubett, C. P., and duPreez, S. F. M.,** The determination of phytic acid and factors which influence its hydrolysis in bread, *Proc. Nutr. S. Africa,* 2, 69, 1961.

56. **Reddy, N. R., Balakrishnan, C. V., and Salunkhe, D. K.,** Phytate phosphorus and mineral changes during germination and cooking of black gram (*Phaseolus mungo* L.) seeds, *J. Food Sci.,* 43, 540, 1978.
57. **Cilliers, J. J. L. and Van Niekerk, P. J.,** LC determination of phytic acid in food by post-column colorimetric detection, *J. Agric. Food Chem.,* 34, 680, 1986.
58. **Earley, E. B.,** Determining phytin phosphorus, *Ind. Eng. Chem. (Anal. Ed.),* 16, 389, 1944.
59. **Samotus, B. and Schwimmer, S.,** Indirect method for determination of phytic acid in plant extracts containing reducing substances, *Biochim. Biophys. Acta.,* 57, 389, 1962.
60. **Makower, R. U.,** Extraction and determination of phytic acid in beans *(Phaseolus vulgaris), Cereal Chem.,* 47, 288, 1970.
61. **Magrill, D. S.,** Micro determination of phytate in solution or bound to hydroxyapatite, *Anal. Biochem.,* 49, 522, 1972.
62. **Eklund, A.,** The contents of phytic acid in protein concentrates prepared from niger seed, sunflower seed, rape seed, and poppy seed, *Upsala J. Med. Sci.,* 80, 5, 1975.
63. **Morris, E. R. and Ellis, R.,** Isolation of monoferric phytate from wheat bran and its biological value as an iron source to the rat, *J. Nutr.,* 106, 753, 1976.
64. **Cheryan, M.,** Phytic acid interactions in food systems, *CRC Crit. Rev. Food Sci. Nutr.,* 13, 297, 1980.
65. **Brown, E. C., Heit, M. L., and Ryan, D. E.,** Phytic acid: an analytical investigation, *Can. J. Chem.,* 39, 1290, 1961.
66. **Anderson, G.,** Effect of iron/phosphorus ratio and acid concentration on the precipitation of ferric inositol hexaphosphate, *J. Sci. Food Agric.,* 14, 352, 1963.
67. **Reddy, N. R. and Salunkhe, D. K.,** Interactions between phytate, protein, and minerals in whey fractions of black gram, *J. Food Sci.,* 46, 564, 1981.
68. **Haug, W. and Lantzsch, H. J.,** Sensitive method for the rapid determination of phytate in cereals and cereal products, *J. Sci. Food Agric.,* 34, 1423, 1983.
69. **Heggen, M. H. M. and Reith, J. F.,** Determination of phytic acid by oxidation of its inositol with periodic acid, *Pharm. Weekblad.,* 87, 801, 1948.
70. **Pons, W. A., Stansbury, M. F., and Holfpauir, C. L.,** An analytical system for determining phosphorus compounds in plant materials, *J. Assoc. Off. Anal. Chem.,* 36, 492, 1953.
71. **Oberleas, D.,** Dietary Factors Affecting Zinc Availability, Ph.D. thesis, University of Missouri, Columbia, 1964.
72. **Mohamed, A. I., Perera, P. A. J., and Hafez, Y. S.,** New chromophore for phytic acid determination, *Cereal Chem.,* 63, 475, 1986.
73. **Caldwell, A. G. and Black, C. A.,** Inositol hexaphosphate: I. Quantitative determination in extracts of soils and manures, *Proc. Soil. Sci. Soc. Amer.,* 22, 290, 1958.
74. **Harland, B. F. and Oberleas, D.,** Anion-exchange method for determination of phytate in foods: collaborative study, *J. Assoc. Off. Anal. Chem.,* 69, 667, 1986.
75. **Cosgrove, D. J.,** The determination of myo-inositol hexaphosphate (phytate), *J. Sci. Food Agric.,* 31, 1253, 1980.
76. **Tangendjaja, B., Buckle, K. A., and Wootton, M.,** Dephosphorylation of phytic acid in rice bran, *J. Food Sci.,* 46, 1021, 1981.
77. **Graf, E. and Dintzis, F. R.,** Determination of phytic acid in foods by high-performance liquid chromatography, *J. Agric. Food Chem.,* 30, 1094, 1982.
78. **Fitchett, A. W. and Woodruff, A.,** Determination of polyvalent anions by ion chromatography, *Liq. Chromat. HPLC Mag.,* 1, 48, 1983.
79. **Kikunaga, S., Takahashi, M., and Huzisige, H.,** Accurate and simple measurement of phytic acid contents in cereal grains, *Plant Cell Physiol.,* 26, 1323, 1985.
80. **Ellis, R., Morris, E. R., and Philpot, C.,** Quantitative determination of phytate in the presence of high inorganic phosphate, *Anal. Biochem.,* 77, 537, 1977.
81. **Phillippy, B. Q., Johnston, M. R., Tao, S. H., and Fox, M. R. S.,** Inositol phosphates in processed foods, *J. Food Sci.,* 53, 496, 1988.
82. **Phillippy, B. Q., Johnston, M. R., Tao, S. H., Fox, M. R. S., and White, K. D.,** Inositol phosphate derivatives of phytic acid in food products, *Fed. Proc.,* 45, 375, 1986.

Chapter 6

OCCURRENCE, DISTRIBUTION, CONTENT, AND DIETARY INTAKE OF PHYTATE

I. INTRODUCTION

Phytic acid (myoinositol 1,2,3,5/4,6-hexakis dihydrogen phosphate) widely occurs in plant seeds and/or grains,[1-7] roots and tubers,[2,8] fruits and vegetables,[9] pollen of various plant species,[10,11] and organic soils.[12,13] The phytate fraction of organic soil consists of a mixture of phosphorylated derivatives of myo-, DL-, scyllo-, and neoinositol. Inositol phosphates with fewer than six phosphate groups, such as myoinositol 1,3,4,5,6-pentakisphosphate, have been isolated and identified from the nucleated erythrocytes of birds, turtles and freshwater fish.[14-18]

II. OCCURRENCE

Phytic acid occurs primarily as a salt of mono- and divalent cations in discrete regions of cereal grains and legumes.[19-21] It rapidly accumulates in seeds and/or grains during ripening period,[22-32,173,174,176] accompanied by other storage substances such as starch and lipids. The accumulation site of phytate in cereals and legumes is in the electron-dense aleurone particles or globoids.[20,21,33-36] The electron-dense aleurone particles are located in the aleurone layer of cereals. The aleurone particles are composed of at least two major parts; one is the high phytate-containing particle and the other is the surrounding coat which consists of protein and carbohydrate[35,37] (Figure 1). In rice, these aleurone particles are spherical, about 1 to 3 μm in diameter.[35]

In legumes and many other seeds, the electron-dense globoid crystals are located within the proteinaceous matrix of protein bodies.[38-40] These are present in the cotyledons of legumes and not in the seed coats. Globoid crystals are structurally and chemically distinct areas within the protein bodies.[36] They vary in size and number depending upon the species. For example, Prattley and Stanley[41] reported that isolated soy globoids varied in size (0.1 to 1.0 μm) and were comparatively small in relation to protein bodies (2 to 20 μm). However, some seeds may lack globoid crystals and still contain phytic acid. Lott et al.[39] reported that the protein bodies in peas lack globoid crystals and contain phytic acid within the protein body. Occurrence and biogenesis of globoid crystals within the protein bodies may be controlled by the calcium, magnesium, and potassium content.[40,42] The presence of phytate within the globoid crystals or aleurone particles has been shown for a wide range of cereals (oats,[43] barley,[44,45] wheat,[33] rice,[31,35,46,47] and sorghum[48]) and legumes (peas,[34,39,40] soybeans,[34,41] linseed,[49] peanuts,[50,51] and broad beans[34,52,53]). Phytic acid can range up to 60 to 80% of the dry weight of globoids.[20,54] The chemical composition of phytic acid-rich particles (aleurone particles) of rice, and globoids of cottonseed, soybean, and peanuts, and phytic acid-rich particles of Great Northern beans is presented in Table 1. These were isolated by different methods. The chemical composition of these isolated phytic acid-rich particles is characterized by high phytic acid, potassium, magnesium, and calcium concentrations (Table 1). Lui and Altschul[54] reported that the isolated globoids from cottonseeds had low amounts of protein, carbohydrate, and lipid and 60% and 10% phytic acid and metals (potassium, magnesium, and calcium), respectively. Major components of globoids from peanuts[51] were protein (35.1%), phytic acid (28.0%), and metals (5%). The globoidal fraction accounted for about 50% of the total magnesium and phytic acid and 13% and 80% of potassium and calcium, respectively, in the protein bodies of peanuts. The isolated phytic acid-rich particles of Great Northern beans[55] contained 34.3% protein, 30.0% carbohydrate, 26.6%

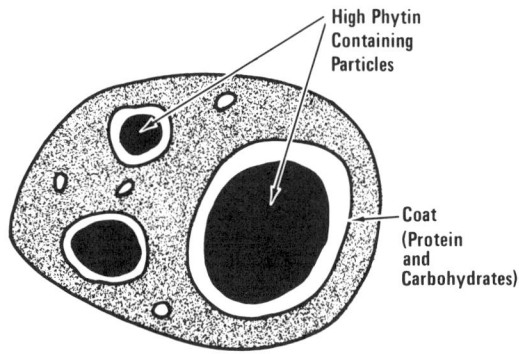

FIGURE 1. Schematic diagram showing an aleurone particle with inclusions, as in rice grains. (From Ogawa, M., Tanaka, K., and Kasai, Z., *Agric. Biol. Chem.*, 39, 695, 1975. With permission.)

TABLE 1
Chemical Composition of Phytic Acid-Containing Particles Isolated From Cottonseed, Peanuts, Soybeans, Great Northern Beans (GNB), and Rice

Composition (weight %)	Globoids of cottonseed	Globoids of peanuts[a]	Globoids of soybeans	Isolated particles of GNB	Isolated particles of	
					Rice bran	Rice embryo
Nitrogen	0.70	—	—	—	—	—
Protein	—	35.10	—	34.30	0.66	0.84
Carbohydrate	1.35	N.S	—	30.00	0.72	0.78
Phytic acid	—	28.00	23.80	26.58	66.68	70.30
Organic phosphorus	13.85	7.30	—	—	—	—
Inositol	13.21	7.00	—	—	—	—
Minerals						
Potassium	6.40	2.00	3.61	0.26	14.55	22.83
Magnesium	1.70	2.50	1.59	2.05	11.97	9.50
Calcium	1.30	0.50	0.86	0.64	0.83	0.73
Moisture	9.71	8.60	—	—	—	—

Note: N.S. indicates result is not significant. For details of isolation, see References 21, 35, 41, 51, 54, 55.

[a] Globoids of peanuts were obtained after centrifugation at 20,000 × g.

phytic acid, and 3% metals. Ogawa et al.[35] reported that over 90% of the compounds of the isolated particles of rice bran were phytic acid, potassium, and magnesium (Table 1). The isolated particles of rice bran and embryo contained small amounts of protein and carbohydrate.[21]

Phytic acid occurs primarily as the potassium-magnesium salt in rice and rice scutellum (part of the embryo),[30,35,56] wheat,[33] broad beans,[34] and sesame,[57] and as the calcium-magnesium-potassium salt in soybeans[34,41,58] and Great Northern beans.[55] However, some of the recent studies have indicated that the phytic acid may be present in water-soluble as well as water-insoluble forms as a salt of metals and proteins.[41] Prattley and Stanley[41] concluded that the phytic acid is located within the protein bodies of soybeans as a water-soluble salt (soluble protein-phytate salt), with significant amounts (estimated at 10 to 15%, depending upon the distribution of globoids) specifically deposited in globoid inclusions in the water-insoluble form. Great Northern beans are reported to contain phytic acid in the water-soluble and water-insoluble

TABLE 2
Phytate Content of Soybean Seeds at Three Stages of Maturity (Planting Date June 16, 1980)[60]

Stage of maturity	Harvest date	Phytate P(%)	Phytate (%)
1	September 20	0.24	0.87
2	September 26	0.30	1.08
3 (mature)	October 3	0.36	1.26

TABLE 3
Changes in Phytate Content During Winged Bean Seed Development[61]

Winged bean origin	Stage of maturity	Phytate P (mg/100g)	Phytate (mg/100g)
Sri Lanka	I	19	67.4
	II	59	209.2
	III	122	432.6
	IV	169	599.2
	Mature	181	641.8
Nigeria	I	19	67.4
	II	53	187.9
	III	105	372.3
	IV	130	461.0
	Mature	179	634.7
Indonesia	I	21	74.5
	II	56	198.6
	III	110	390.1
	IV	134	475.2
	Mature	202	716.3

Note: The stages of maturity are as follows: I=40 days, II=50 days, III=60 days, IV=70 days, and maturity = 80 to 85 days after flowering.

forms as a salt of metals and protein.[55] In a separate study, Reddy et al.[59] found that phytic acid is present as a water-soluble salt with a molecular weight of less than 1,000 daltons and water-soluble complex with molecular weight more than 1,000 daltons in Great Northern beans. However, the exact water-soluble forms in which phytic acid occurs in Great Northern beans have not been identified. Lott et al.[39] suggested that most of the phytate in peas is water soluble and present as potassium phytate. There is a need for further studies to identify the chemical form(s) in which phytic acid occurs in cereals and legumes.

III. PHYTATE DISTRIBUTION AND CONTENT

In cereals and legumes, phytate accumulates during seed development and reaches its highest level at seed maturity.[21,27,28,60-63] Yao et al.[60] reported that during soybean maturation the phytate content increases from 0.87% to 1.26% on a dry weight basis (Table 2). The phytate content increased from 1.0 to 2.2 mg between Stage 1 and Stage 3, when calculated on a per bean basis. Welch et al.[63] found that during maturation of peas the phytate content increased from 0.16% to 1.23%. In winged beans, the proportion of phytate increased at all stages of seed maturity (Table 3).[61]

TABLE 4
Phytate Concentration and Distribution in Morphological Components of Cereals[64,65]

Cereal	Morphological component	Phytate phosphorus (%)	Phytate[a] (%)	Distribution[b] (%)
Corn	Commercial hybrid	0.25	0.89	—
	Endosperm	0.01	0.04	3.20
	Germ	1.80	6.39	88.00
	Hull	0.02	0.07	0.04
Corn	High lysine	0.27	0.96	—
	Endosperm	0.01	0.04	3.00
	Germ	1.61	5.72	88.90
	Hull	0.07	0.25	1.50
Wheat	Soft	0.32	1.14	—
	Endosperm	0.001	0.004	2.20
	Germ	1.10	3.91	12.90
	Aleurone layer	1.16	4.12	87.10
Rice	Brown	0.25	0.89	—
	Endosperm	0.004	0.01	1.20
	Germ	0.98	3.48	7.60
	Pericarp	0.95	3.37	80.00
Pearl millet	Whole	0.25	0.89	—
	Endosperm	0.09	0.32	—
	Germ	0.75	2.66	—
	Bran	0.28	0.99	—

[a] Phytate content is calculated by assuming that it contains 28.20% phosphorus.
[b] Percentage of phytate in the component part.

A. PHYTATE DISTRIBUTION

In cereals, phytate is associated with specific components within the grain and can be preferentially separated with those components. The phytate concentration in morphological components or parts of cereal grains is presented in Table 4. The endosperm of wheat and rice kernels is almost devoid of phytate, as it is concentrated in the germ and aleurone layers (pericarp) of the cells of the kernel. Corn differs from most other cereals since 88.0% of phytate is concentrated in the germ portion of the kernel.[64] Corn endosperm has small amounts (3.2%) of phytate. Rice and wheat germ contain appreciable amounts of phytate but a major portion of phytate is in the aleurone layers (pericarp). Of the total phytate, 84.0 to 88.0% was reported to be present in the bran part of the rice.[66] In pearl millets, the majority of the phytate appears to be present in germ and bran fractions.[65] However, the distribution in pearl millets is not determined. Frolich and Nyman[177] reported that more than 90% of the phytate was found in the soluble fiber fraction of oats.

In legumes, phytate is distributed throughout the cotyledon and located within the subcellular inclusions of protein bodies.[36,42] Ferguson and Bollard[67] found that 99% of the phytate in dry peas was in the cotyledons and 1% in the embryo axis. Phytate phosphorus represents about 65% of the total phosphorus in the cotyledons and 20% of the total phosphorus in the embryo axis. Beal and Mehta[68] reported that more than 88% of the phytate is present in pea cotyledon (Table 5). The hull or seed coat fraction contains little or no phytate.

B. PHYTATE CONTENT
1. Cereals and Cereal Products

The phytate phosphorus and phytate content of cereals, cereal products, and cereal-based foods is presented in Table 6. The amount of phytate varies from 0.14 to 2.22% in cereals, 0.08

TABLE 5
Phytate Concentration and Distribution in Morphological Components of Peas[68]

Morphological component	Phytate P(%)	Phytate[a] (%)	Distribution[b] (%)
Whole pea	0.22	0.79	—
Cotyledon	0.22	0.78	88.70
Germ	0.35	1.23	2.50
Hull	0.002	0.01	0.10

[a] Phytate content is calculated by assuming that it contains 28.20% phosphorus.
[b] Percentage of phytate in the component part.

TABLE 6
Phytate Phosphorus and Phytate Content of Cereals, Cereal Products, and Cereal-Based Foods

Cereals/cereal products	Phytate phosphorus (%)	Phytate (%)	Ref.
Cereals			
Wheat	0.11—0.38	0.39—1.35	69—73, 178
Hard wheat	0.24	0.84	74
Soft wheat	0.27—0.32	0.94—1.13	64, 75
Durum wheat	0.25—0.33	0.88—1.16	76
Corn	0.23—0.63	0.83—2.22	71, 72, 77
Corn (high lysine)	0.28	0.99	77
Popcorn (unpopped)	0.17	0.62	179
Triticale	0.14—0.53	0.50—1.89	75, 78
Oat	0.20—0.33	0.70—1.16	70, 71, 73, 79, 80, 177
Barley	0.21—0.33	0.75—1.16	9, 69—71
Rye	0.15—0.41	0.54—1.46	69, 71, 81, 82
Sorghum	0.26—0.38	0.91—1.35	83-85
Sorghum (low tannin)	0.16	0.57	86
Sorghum (high tannin)	0.27	0.96	86
Proso millets	0.17—0.47	0.60—1.67	87
Pearl millets	0.05—0.09	0.18—0.31	65
Ragi	0.15—0.19	0.55—0.67	184
Wild rice	0.62	2.20	71
Brown rice (long-grain)	0.24—0.25	0.84—0.89	72, 77
Polished rice (long-grain)	0.10—0.14	0.34—0.50	9, 88
Polished rice (medium-grain)	0.04—0.05	0.14—0.19	72, 89
Polished rice (short grain)	0.04	0.14	89
White rice (enriched)	0.06	0.23	90
Glutinous rice (Vietnam)	0.06	0.23	91
Cereal milled fractions and protein products			
Wheat flour	0.07—0.39	0.25—1.37	93, 170
Wheat flour (India)	0.14—0.16	0.50—0.55	92
Wheat bran	0.80—1.44	2.85—5.11	71, 79, 94, 169, 170, 178
Wheat gluten	0.60	2.13	97, 98
Wheat germ	0.023	0.083	91
Corn meal	0.23—0.30	0.81—1.07	179
Corn germ	0.55	1.94	179
Rye flour	0.30	1.08	179
Triticale flour	0.19	0.67	179
Oatmeal	0.29—0.68	1.04—2.40	71, 179
Oat white flour	0.11	0.40	177

TABLE 6 (continued)
Phytate Phosphorus and Phytate Content of Cereals, Cereal Products, and Cereal-Based Foods

Cereals/cereal products	Phytate phosphorus (%)	Phytate (%)	Ref.
Oat bran	0.17	0.60	177
Rice bran	0.73—1.54	2.59—5.46	178
Wheat protein concentrate	0.53—0.76	1.88—2.70	95, 96
Cereal-Based Foods			
Wheat breads			
Whole wheat breads	0.12—0.23	0.43—0.82	71, 72, 79, 99—101, 179
Whole wheat bread (flat)	0.12—0.23	0.43—0.83	72, 101
Wheat bread (Roman meal)	0.11—0.13	0.28—0.45	90, 101
Wheat breads (Rainbow)	0.08	0.29	101
Honey wheat berry bread	0.21	0.73	101
Wheat bread (Dillon's)	0.28	1.00	101
Whole wheat bread (Safeway)	0.30	1.05	101
Wheat bread (high fiber)	0.10	0.37	179
White breads			
White bread	0.01—0.04	0.03—0.13	71, 99, 100
White bread (Rainbow)	0.03	0.11	101
White bread (Mrs. Wright's)	0.06	0.23	101
White bread (Sky Lark)	0.06	0.23	101
White bread (high fiber)	0.03	0.12	179
Other breads			
Corn bread	0.38	1.36	100
Rye bread	0.12	0.41	100
French bread	0.01	0.03	100
Raisin bread	0.03	0.10	100
Pumpernickel bread	0.05	0.16	100
Plain hamburger bun	0.03	0.12	179
Pita bread	0.05	0.16	179
Indian chapatis (flat bread)	0.07—0.16	0.25—0.56	92, 102
Norwegian flat (Kauli)	0.19	0.68	179
Iranian Flat breads			
Bazari (leavened)	0.33	1.17	107, 108
Sangak (leavened)	0.39	1.38	107, 108
Tanok (unleavened)	0.68	2.41	107, 108
Pakistani Flat breads and cereal foods			
Flat breads	0.18—0.19	0.67—0.68	103
Roti	0.13—0.15	0.47—0.52	103
Nan	0.01	0.04	103
Wheat porridge	0.05	0.16	103
Puri	0.01	0.03—0.04	103
Italian cereal foods			
Wheat bran	0.96—0.98	3.41—3.47	104
Breads	0.02—0.07	0.06—0.26	104
Crackers	0.10—0.16	0.37—0.58	104
Biscuits	0.03—0.30	0.11—1.05	104
Other foods			
Wheat bran muffins	0.22—0.36	0.77—1.27	94, 169, 177
English muffins	0.03	0.12	179
Wheat germ pancakes	0.50	1.76	179
Doughnut cake, sugar-coated	0.14	0.48	179
Brown rice, parboiled	0.44	1.60	105

TABLE 6 (continued)
Phytate Phosphorus and Phytate Content of Cereals, Cereal Products, and Cereal-Based Foods

Cereals/cereal products	Phytate phosphorus (%)	Phytate (%)	Ref.
Barley, pearl boiled	0.15	0.54	179
Corn chips	0.07—0.19	0.24—0.66	179
Crackers			
Animal	0.03	0.09	179
Graham	0.05	0.19	179
Ritz	0.03	0.11	179
Saltines	0.05	0.18	179
Wheat thins	0.10	0.34	179
Ready-to-eat cereals			
Wheat cereal (40% bran)	0.32	1.12	179
Wheat cereal (100% bran)	0.93	3.29	179
Wheat cereal (Bran Chex)	0.39	1.39	179
Wheat cereal (Bran Flakes)	0.31	1.10	179
Wheat cereal (Frosted Flakes)	0.02	0.06	179
Wheat cereal (Grape Nuts)	0.16	0.56	179
Wheat cereal (Honey Smacks)	0.05	0.19	179
Wheat cereal (Post Toasties)	0.02	0.08	179
Wheat cereal (Raisin Bran)	0.20—0.52	0.72—1.83	71, 179
Wheat cereal (Shredded Wheat)	0.43	1.53	71, 179
Wheat cereal (Special K)	0.20	0.69	71, 179
Wheat cereal (Super Sugar Crisps)	0.08	0.28	179
Wheat cereal (Wheaties)	0.43	1.52	179
Corn cereal (corn bran)	0.07	0.24	179
Corn cereal (Corn Flakes)	0.01—0.03	0.05—0.09	71, 99, 179
Corn cereal (Corn Pops)	0.03	0.10	179
Mixed grain cereal (Apple Bits)	0.15	0.52	179
Mixed grain cereal (Apple Jacks)	0.05	0.18	179
Mixed grain cereal (Froot Loops)	0.05	0.17	179
Mixed grain cereal (Product 19)	0.08—0.14	0.28—0.50	71, 179
Mixed grain cereal (Special K)	0.08	0.28	179
Mixed grain cereal (Team)	0.07	0.24	179
Rice cereal (Cocoa Krispies)	0.04	0.14	179
Rice cereal (Cocoa Pebbles)	0.05	0.19	179
Rice cereal (Fruity Pebbles)	0.04	0.15	179
Rice cereal (Rice Krispies)	0.05—0.07	0.18—0.24	71, 94, 179
Quick oats (Quaker)	0.28	0.99—1.00	94
Old fashioned oats (Quaker)	0.18	0.64	106
Instant oats with bran and raisins (Quaker)	0.28	0.99	106
Quick oats (Safeway)	0.29	1.03	106
Infant cereals			
Infant cereal	0.39	1.38	71
Barley, infant cereal	0.28	1.00	179
Rice cereal (dry form)	0.28	0.98	179
Mixed grain infant cereal (instant, dry form)	0.23	0.81	179

to 5.46% in cereal-milled fractions and protein products, 0.05 to 3.29% in ready-to-eat cereal products, 0.81 to 1.38% in infant cereals, 0.09 to 0.34% in crackers, and 0.03 to 2.41% in cereal-based foods (wheat bread, white breads, other types of breads, flat breads, and other foods). Among all cereals, pearl millet and various types of polished rice contain low amounts of phytate. Some of the ready-to-eat cereals (100% wheat bran cereal, shredded wheat, wheaties,

TABLE 7
Phytate Content in Triticale, Wheat, and Rye and Their Milled Fractions,[a] Bran, and Flour[75]

	Phytate (%)		
Cultivar	Whole grain	Bran	Flour
Triticales			
6TA 131	0.81[b]	0.47[b]	0.25[b]
6TA 203	1.08	0.61	0.19
6TA 204	0.52	0.47	0.33
6TA 205	0.49	0.52	0.19
6TA 385	1.47	0.67	0.41
6TA 418	0.59	0.61	0.33
6TA 419	1.25	0.78	0.48
6TA 514	0.83	0.32	0.20
72 S	0.52	0.47	0.36
NB 69150	1.27	0.61	0.45
K.S. Bulk	1.89	0.47	0.37
Mark IV	0.88	0.31	0.33
8X Triticale	1.35	0.98	0.18
Wheat			
Arthur 71	0.94	0.85	0.20
Rye			
Abruzzi	0.97	0.61	0.33

[a] Milled fractions obtained after milling with Brabender Quadrumat Junior Mill.
[b] Data expressed on dry weight basis.

raisin bran wheat cereal, and infant cereals) have high amounts of phytic acid (Table 6). White breads, other breads (French, rye, raisin, and pumpernickel), and Pakistani flat breads have a low phytate content compared to wheat breads and Iranian flat breads. The differences in phytate content of various breads could be due to (1) degree of fermentation during preparation and (2) use of different extraction rate flours in preparing breads. For instance, Iranian flat breads are prepared using different extraction rate flours, which results in breads with high phytate content.[107,108] During fermentation of bread dough, some of the phytate in the dough will be hydrolyzed by wheat and bacterial phytase, which may result in a reduced phytate content, especially breads with white flours. Since most of the phytate in cereals is located in the aleurone layers (bran), milling of cereals and subsequent separation of bran results in a significant reduction of phytate in flours.[75] The bran fractions of triticale, wheat, and rye have higher amounts of phytate than the corresponding flours (Table 7).

Phytate phosphorus accounts for the major portion (more than 80%) of total phosphorus in cereals and cereal products. Of the total phosphorus, phytate phosphorus represents 81.0% in brown rice, 51.0 to 61.0% in polished grain rice,[88,89] 60.0 to 80.0% in wheat,[36,64,70,74,178] 66.0 to 70.0% in barley,[36,70] 48.7 to 70.9% in oats,[36,70,177] 38.0 to 46.0% in rye,[36,70,81] 18.0 to 73.0% in triticale,[78,81] 71.0 to 88.0% in corn [36,77,178] 87.1% in high lysine corn,[36] 72.0 to 90.5% in sorghum,[85] 88.9% in high tannin sorghum,[86] 72.0 to 78.0% in ragi,[184] 73.0 to 85.7% in rice bran,[178] 71.5 to 89.0% in wheat bran,[106,178] 54.0% in oat bran,[177] 80.0% in oat white flour,[177] 84.0% in Quaker instant oats with bran and raisins,[106] 34.0% in Quaker old fashioned oats,[106] 15.0 to 33.0% in white breads,[101,106] 55.0% in brown bread,[101] and 38.0 to 66.0% in whole wheat breads.[101]

2. Legumes and Legume Products

Phytate content ranges from 0.22 to 9.15% in whole beans, 0.58 to 4.20% in bean flours and bean protein products, 0.05 to 5.20% in bean-based foods, and 0.004 to 0.03% in crude soybean oil (Table 8). Among legumes, dolique beans contain the highest amount of phytate, i.e., 5.92 to 9.15%. Crude soybean oil and soy milk appear to contain the lowest amount of phytate. The wide variations reported for phytate content within the same type of bean may be due to differences in cultivars. Because most of the phytate in beans is distributed in the cotyledons, removal of the hull or seed coat could lead to a higher phytate content of beans[68,102,119,129,136] (Table 9).

In beans, phytate phosphorus constitutes a major portion of the total phosphorus. For example, of the total phosphorus, phytate phosphorus accounts for 50.0 to 70.0% in soybeans,[36,70,116] 27.0 to 87.0% in lentils,[116] 40.0 to 95.0% in chickpeas,[116,125] 39.5 to 95.0% in broad beans,[36,116] 37.0% in peas,[116] 76.0% in pigeon peas,[116] 70.0 to 87.0% in linseed,[116] 31.0 to 60.0% in lima beans,[130] 69.0% in green gram,[74,125] 37.0 to 54.0% in dolique beans,[116] 29.8 to 50.0% in cowpeas,[126] 79.0% in black gram,[129,152] 57.0 to 81.0% in navy beans,[132] 68.0 to 72.0% in red kidney beans,[132] 55.0 to 80.0% in Great Northern beans,[55,132] 57.0% in peanuts,[36] 44.0 to 73.0% in winged beans,[61,109] 70.0% in California Small White beans,[74] 54.7% in lupine,[178] 77.0% in tempe, and 94.5% in tofu,[140] 87.0% in defatted soy flour,[140] and 62.0% in soy protein isolate.[140]

C. EFFECTS OF ENVIRONMENTAL AND OTHER FACTORS ON THE PHYTATE CONTENT OF CEREALS AND LEGUMES

Environmental fluctuations, locations, irrigation conditions, type of soil, various fertilizer applications, and year during which a cultivar or variety is grown influences the phytate content of seeds. Bassiri and Nahapetian[153] observed that wheat varieties grown under dry land conditions had lower concentrations of phytate than the ones grown under irrigated conditions. Nahapetian and Bassiri,[154] Singh and Reddy,[81] Singh and Sedeh,[78] Miller et al.,[155,156] and Simwemba et al.[65] reported variations in the phytate content of triticales, wheat, rye, oats, and pearl millet grown at different locations and in different years. A variation in phytate content of navy beans was observed by Proctor and Watts[157] as a result of cultivar and location effects. Griffiths and Thomas[158] reported that the phytate phosphorus content of broad beans was increased substantially (from 39.5 to 57.7% when calculated as percent of total phosphorus) when they were grown under glasshouse conditions compared to field conditions. Application of different regimes of fertilizers (nitrogen and phosphorus) to field crops during their growth is reported to increase in phytate content of their seeds and/or grains.[80,156,159,160,172-175]

IV. DIETARY INTAKE OF PHYTATE

It is difficult to estimate the daily intake of phytate for populations around the world. Harland and Peterson[161] suggested that the average American (weighing 75 kg) consumes about 750 mg phytate per day. In general, vegetarians consume a higher amount of phytate than nonvegetarians.[162] Phytate intake also varies with the season. Recently, Ellis et al.[162] reported that phytate intake of omnivorous, self selected diets varied from 585 mg/day in spring to 734 mg/day in winter for females and from 781 mg/day in spring to 762 mg/day in winter for males. Cereal may be the major source of dietary phytate for American men consuming an omnivorous diet. The daily intake of phytate ranges from 615 to 5770 mg in lacto-ovovegetarian diets of Trappist Monks.[161] Murphy and Calloway[181] reported phytate intakes of 395 mg/day by 996 non-pregnant, non-lactating women, aged 18 to 24 years, in the second National Health and Nutrition Examination Survey (NHANES II).

Davies[163] estimated that the phytate intake in the U.K. ranges from 600 to 800 mg/day. About 70% of this phytate intake is derived from cereal products, 20% from fruits, and the remainder from vegetables and nuts. On the other hand, Swedish people appear to consume diets containing very low levels of phytate.[182] In 13 Italian diets, daily phytate intake varied from a maximum of 1367 mg to a minimum of 112 mg.[164] This large variation is due to the consumption of a variety of refined cereal and pasta products. Indian diets may also provide high phytate intakes because

TABLE 8
Phytate Phosphorus and Phytate Content of Whole Beans, Bean Flours and Protein Products, and Bean-Based Foods

Bean/bean products	Phytate phosphorus (%)	Phytate (%)	Ref.
Whole beans			
Winged beans	0.18—0.75	0.63—2.67	61, 109—112, 185
Soybeans	0.28—0.63	1.00—2.22	70, 71, 77, 83, 91, 93, 99, 113, 114
Broad beans (faba beans)	0.15—0.50	0.54—1.77	91, 115—120, 180
Broad beans, boiled	0.03	0.11	179
Dolique beans	0.17—0.26	5.92—9.15	116
Peas	0.06—0.33	0.22—1.22	9, 63, 91, 115, 116, 179, 180
Dwarf Grey pea	0.11	0.40	121
Early Alaskan pea	0.19	0.67	121
Pigeon pea	0.20—1.97	0.71—7.00	102, 116, 122, 123
Pigeon pea (early maturing)	0.35—0.46	1.25—1.64	124
Pigeon pea (medium maturing)	0.31—0.45	1.10—1.60	124
Pigeon pea (late maturing)	0.31—0.46	1.10—1.63	124
Lentil	0.08—0.30	0.27—1.05	9, 91, 93, 99, 102, 115, 116, 171
Linseed	0.61—0.79	2.15—2.78	116
Peanuts, Spanish	0.53	1.88	90
Peanuts	0.50	1.76	71
Chickpeas	0.08—0.35	0.28—1.26	9, 99, 102, 115, 116, 125, 179
Cowpeas	0.12—0.41	0.44—1.45	114, 120, 125—127
Moth bean	0.24—0.26	0.85—0.92	128
Green gram	0.19—0.31	0.67—1.10	102, 123, 129
Black gram	0.20—0.41	0.72—1.46	102, 123, 129
Lima beans	0.15—0.71	0.53—2.52	71, 72, 74, 120, 130
Lima beans (immature, raw)	0.20	0.70	179
Blackeye peas	0.26—0.39	0.91—1.38	99, 102, 179
Lupine	0.29—0.30	1.02—1.06	178
Kidney beans	0.25—0.44	0.89—1.57	99, 102
California Small White beans	0.07—0.29	0.26—1.03	72, 74, 120, 131
Pinto beans	0.17—0.67	0.61—2.38	90, 120, 132—134, 179
Navy beans	0.21—0.50	0.74—1.78	71, 132, 179
Great Northern beans	0.16—0.58	0.56—2.70	55, 132, 135—137
Small white beans	0.15—0.33	0.55—1.16	132, 136
Red Mexican beans	0.15—0.31	0.54—1.10	132
Red kidney beans	0.34—0.58	1.20—2.06	132, 133
Small red beans	0.58	2.07	136
Sanilac beans	0.78	2.75	134
Cranberry beans	0.74	2.63	134
Viva Pink beans	0.61	2.16	134
Black Beauty beans	0.29—0.83	1.04—2.93	120, 136
Light red kidney beans	0.34—0.74	1.20—2.63	131, 136
Dark red kidney beans	0.81	2.86	136
Yellow peas, split	0.15—0.47	0.54—1.68	105, 171, 179
Green peas, split	0.24	0.85	90
Bean flours and protein products			
Soy flour, full-fat	0.35—0.37	1.24—1.30	139, 140
Soy flour, defatted	0.43—0.63	1.52—2.25	94, 105, 140
Soy flakes, defatted	0.43—0.52	1.52—1.84	77, 90, 91
Soy protein concentrate	0.35—0.61	1.24—2.17	94, 96, 139, 141, 142, 169
Soy protein isolate	0.39—0.60	1.40—2.11	79, 94, 96, 140, 142
Soy beverage	0.35	1.24	139
Soy bean oil (crude)	0.001—0.009	0.004—0.03	143
Peanut flour	0.42—0.55	1.50—1.94	79

TABLE 8 (continued)
Phytate Phosphorus and Phytate Content of Whole Beans, Bean Flours and Protein Products, and Bean-Based Foods

Beans/bean products	Phytate phosphorus (%)	Phytate (%)	Ref.
Peanut meal, defatted	0.48	1.70	50
Linseed meal, defatted	1.18	4.20	144
Pea protein concentrate	0.16—0.54	0.58—1.90	93, 171, 180
Faba bean protein concentrate	0.42—0.96	1.48—3.39	93, 145
Lentil protein concentrate	0.29	1.02	171
Bean-based foods			
Soy milk	0.014—0.03	0.05—0.11	146
Tempe	0.27—0.30	0.67—1.08	113, 140, 147
Tofu	0.55—0.82	1.96—2.90	83, 140
Calcium-tofu	0.45—0.54	1.60—1.91	32
Magnesium-tofu	0.42—0.50	1.49—1.77	32
Idli	0.15	0.54	148
Khaman	1.45	5.20	149
Dhokla	0.51	1.80	150
Oncom	0.20	0.70	151
Peanuts, toasted and salted	0.28	1.00	179
Chicken analog	0.08	0.27	141
Ham analog	0.03	0.12	141
TVP pork	0.40	1.42	141
TVP bacon	0.27	0.95	141
TVP ham	0.36	1.26	141
TVP beef	0.38	1.36	141
TVP beef chunks	0.38	1.36	141

Note: TVP = texturized vegetable (soybean) protein.

TABLE 9
Phytate Content of Whole Beans and Bean Cotyledons[68,102,115,119,120,136,180]

Cultivar	Phytate (%) Whole beans	Phytate (%) Bean cotyledons
U.S. dry beans		
Great Northern	2.04	3.26
Small White	1.16	1.63
Sanilac	2.75	2.94
Cranberry	2.63	3.39
Viva Pink	2.16	2.91
Pinto	2.38	2.56
Light red kidney	2.63	3.47
Dark red kidney	2.86	3.67
Small Red	2.07	3.05
Black Beauty	2.93	3.61
Other beans		
Chickpeas	0.56	1.05
Black gram	1.46	1.70
Broad beans	0.86—1.62	0.99—1.90
Lentils	0.44—0.50	0.49—0.53
Lupine	0.30—0.35	0.18—0.24
Peas	0.79	0.78

they consist mainly of cereals and beans. Bindra et al.[165] indicated that the Punjabi diets contain higher phytate levels than the levels reported for both omnivorous and vegetarian American diets. The Punjabi diet consists mainly of chapatis (prepared from 100% extraction wheat flour) and beans. Preparing of chapatis does not involve leavening; therefore the phytate is not destroyed before consumption. An average Nigerian may consume as much as 2000 to 2200 mg of phytate per day using a typical menu of kidney bean balls, rice, plantains, yams, gari, and pudding.[166,167] This is three times the estimated intake of 750 mg in the North American population. Middle East inhabitants also consume very high amounts of phytate in their diets. Recently, Wise et al.[183] conducted a study for estimating mean daily phytate intakes in male and female students consuming different diets. They found that the mean phytate intakes were lower for females (141 mg) compared to males (237 mg). The high intake of phytate may have unfavorable effects on the availability of minerals.[168] For proper evaluation of the effect of phytate on mineral status, one needs to know the phytate content of food and also the intakes of calcium, magnesium, zinc, and fiber in the diets.[181] Harland and Oberleas[179] reported the phytate content for various foods and drinks in three sets of units: mg phytate per serving, mg phytate per 100 g edible portion, and mg phytate per 100 g dry weight of the material. The phytate intake in a mixed or complex diet can be calculated by using these three sets of phytate units. However, the total daily phytate intake may be inadequate for predicting its effects on mineral availability, because the binding of minerals to phytate depends on their relative concentrations in each meal and in digesta passing through the small intestine.[183] The effect of phytate on the bioavailability of minerals is discussed in a separate chapter.

REFERENCES

1. **Posternak, S.,** Sur un nouveau principle phospho-organic d'argine vegetale, *C.R. Soc. Biol.,* 55, 1190, 1903.
2. **Rose, A. R.,** A resumé of the literature on inosite phosphoric acid, with special reference to the relation of that substance to plants, *Biochem. Bull.,* 2, 21, 1912.
3. **Averill, H. P. and King, C. G.,** The phytin content of foodstuffs, *J. Am. Chem. Soc.,* 48, 724, 1926.
4. **Belavady, B. and Banerjee, S.,** Studies on the effect of germination on the phosphorus values of some common Indian pulses, *Food Res.,* 18, 223, 1953.
5. **O'Dell, B. L.,** Effect of soy protein on trace mineral bioavailability, in *Soy Protein and Human Nutrition,* Wilcke, H. L., Hopkins, D. T., and Waggle, D. H., Eds., Academic Press, New York, 1979, 187.
6. **Oberleas, D.,** Phytate content in cereals and legumes and methods of determination, *Cereal Foods World,* 28, 352, 1981.
7. **Graf, E.,** Applications of phytic acid, *J. Am. Oil Chem. Soc.,* 60, 1861, 1983.
8. **McCance, R. A. and Widdowson, E. M.,** Phytin in human nutrition, *Biochem. J.* 29B, 2694, 1935.
9. **Larbi, A. and M'barek, E.,** Dietary fiber and phytic acid levels in the major food items consumed in Morocco, *Nutr. Rep. Intern.,* 31, 469, 1985.
10. **Jackson, J. F., Jones, G., and Linskens, H. F.,** Phytic acid in pollen, *Phytochemistry,* 21, 1255, 1982.
11. **Scott, J. J. and Loewus, F. A.,** Phytate metabolism in plants, in *Phytic Acid: Chemistry and Applications,* Graf. E., Ed., Pilatus Press, Minneapolis, 1986, 23.
12. **Dyer, W. J., Wrenshall, C. L., and Smith, G. R.,** The isolation of phytin from soil, *Science,* 91, 319, 1940.
13. **Caldwell, A. G. and Black, C. A.,** Inositol hexaphosphate. III. Content in soils, *Proc. Soil Sci. Soc. Am.,* 22, 296, 1958.
14. **Rapaport, S.,** Phytic acid in avian erythrocytes, *J. Biol. Chem.,* 135, 403, 1940.
15. **Rapaport, S. and Guest, G. M.,** Distribution of acid-soluble phosphorus in the blood cells of various vertebrates, *J. Biol. Chem.,* 138, 269, 1941.
16. **Oshima, M., Taylor, T. G., and Williams, A.,** Variations in the concentration of phytic acid in the blood of the domestic fowl, *Biochem. J.,* 92, 42, 1964.
17. **Johnson, L. F. and Tate, M. E.,** Structure of phytic acids, *Can. J. Chem.,* 47, 63, 1969.

18. **Isaacks, R. E., Kim, H. D., Bartlett, G. R., and Harkness, D. R.,** Inositol pentaphosphate in erythrocytes of a fresh water fish, peraracu *(Arapaima gigas), Life Sci.,* 20, 987, 1977.
19. **Cosgrove, D. J.,** The chemistry and biochemistry of inositol polyphosphates, *Rev. Pure Appl. Chem.,* 16, 209, 1966.
20. **Sobolev, A. M.,** On the state of phytin in the aleurone grains of mature and germinating seeds, *Soviet-Plant Physiol.,* 13, 177, 1966.
21. **Tanaka, K. and Kasai, Z.,** Phytic acid in rice grains, in *Antinutrients and Natural Toxicants in Foods,* Ory, R. L., Ed., Food and Nutrition Press, Westport, 1981, 239.
22. **Asada, K. and Kasai, Z.,** Formation of phytin and its role in the ripening process of rice plant, *Mem. Res. Inst. Food Sci., Kyoto Univ.,* 18, 32, 1959.
23. **Asada, K. and Kasai, Z.,** Formation of myo-inositol and phytin in ripening rice grains, *Plant Cell Physiol.,* 3, 397, 1962.
24. **Verma, S. C. and Lal, B. M.,** Physiology of bengal gram seed. II. Changes in phosphorus compounds during ripening of the seed, *J. Sci. Food Agric.,* 17, 43, 1966.
25. **Asada, K., Tanaka, K., and Kasai, Z.,** Formation of phytic acid in cereal grains, *Ann. N.Y. Acad. Sci.,* 16, 801, 1969.
26. **Makower, R. V.,** Changes in phytic acid and acid-soluble phosphorus in maturing pinto beans, *J. Sci. Food Agric.,* 20, 82, 1969.
27. **Abernethy, R. H., Paulsen, G. M., and Ellis, R., Jr.,** Relationship among phytic acid, phosphorus, and zinc during maturation of winter wheat, *J. Agric. Food Chem.,* 21, 282, 1973.
28. **Nahapetian, A. and Bassiri, A.,** Changes in concentrations and interrelationships of phytate, phosphorus, magnesium, calcium, and zinc in wheat during maturation, *J. Agric. Food Chem.,* 23, 1179, 1975.
29. **Welch, R. M. and VanCampen, R.,** Iron availability to rats from soybeans, *J. Nutr.,* 105, 253, 1975.
30. **Ogawa, M., Tanaka, K., and Kasai, Z.,** Phytic acid formation in dissected ripening rice grains, *Agric. Biol. Chem.,* 43, 2211, 1979.
31. **Ogawa, M., Tanaka, K., and Kasai, Z.,** Energy dispersive x-ray analysis of phytin globoids in aleurone particles of developing rice grains, *Soil Sci. Plant Nutr.,* 25, 437, 1979.
32. **Forbes, R. M., Parker, H. M., Kondo, H., and Erdman, J. W., Jr.,** Availability to rats of zinc in green and mature soybeans, *Nutr. Res.,* 3, 699, 1983.
33. **Tanaka, K., Yoshida, T., and Kasai, Z.,** Radioautographic demonstration of the accumulation site of phytic acid in rice and wheat grains, *Plant Cell Physiol.,* 15, 147, 1974.
34. **Lott, J. N. A. and Buttrose, M. K.,** Globoids in protein bodies of legume seed cotyledons, *Aust. J. Plant Physiol.,* 5, 89, 1978.
35. **Ogawa, M., Tanaka, K., and Kasai, Z.,** Isolation of high phytin containing particles from rice grains using an aqueous polymer two phase system, *Agric. Biol. Chem.,* 39, 695, 1975.
36. **Lott, J. N. A.,** Accumulation of seed reserves of phosphorus and other minerals, in *Seed Physiology,* Vol. 1, Murray, D. R., Ed., Academic Press, New York, 1984, 139.
37. **Ogawa, M., Tanaka, K., and Kasai, Z.,** Note on the phytin-containing particles isolated from rice scutellum, *Cereal Chem.,* 54, 1029, 1977.
38. **Lott, J. N. A.,** Protein bodies, in *Biochemistry of Plants,* Vol. 1, Tolbert, N. E., Ed., Academic Press, New York, 1980, 589.
39. **Lott, J. N. A., Goodchild, D. J., and Craig, S.,** Studies of mineral reserves in pea *(Pisum sativum)* cotyledons using low-water content procedures, *Aust. J. Plant Physiol.,* 11, 459, 1984.
40. **Lott, J. N. A., Randall, P. J., Goodchild, D. J., and Craig, S.,** Occurrence of globoid crystals in cotyledonary protein bodies of *Pisum sativum* as influenced by experimentally induced changes in Mg, Ca, and K contents of seeds, *Aust. J. Plant Physiol.,* 12, 341, 1985.
41. **Prattley, C. A. and Stanley, D. W.,** Protein-phytate interactions in soybeans. I. Localization of phytate in protein bodies, and globoids, *J. Food Biochem.,* 6, 243, 1982.
42. **Lott, J. N. A. and Ockenden, I.,** The fine structure of phytate-rich particles in plants, in *Phytic Acid: Chemistry and Applications,* Graf, E., Ed., Pilatus Press, Minneapolis, 1986, 43.
43. **Buttrose, M. S.,** Manganese and iron in globoid crystals of protein bodies from *Avena* and *Csuarina, Aust. J. Plant Physiol.,* 5, 631, 1978.
44. **Jacobsen, J. V., Knox, R. B., and Pyliotis, N. A.,** The structure and composition of aleurone grains in the barley aleurone layer, *Planta,* 101, 189, 1971.
45. **Liu, D. J. and Pomeranz, Y.,** Distribution of minerals in barley at the cellular level by x-ray analysis, *Cereal Chem.,* 52, 620, 1975.
46. **Tanaka, K., Yoshida, T., Asada, K., and Kasai, Z.,** Subcellular particles isolated from aleurone layer of rice seeds, *Arch. Biochem. Biophys.,* 155, 136, 1973.
47. **Tanaka, K., Ogawa, M., and Kasai, Z.,** The rice scutellum. II. A comparison of scuteller and aleurone electron-dense particles by transmission electron microscopy including energy-dispersive x-ray analysis, *Cereal Chem.,* 54, 684, 1977.

48. **Adams, C. A. and Novellie, L.,** Composition and structure of protein bodies and spherosomes isolated from ungerminated seeds of *Sorghum bicolor* (Linn) Moench, *Plant Physiol.,* 55, 1, 1975.
49. **Poux, N.,** Localisation de l'activité phosphatasique acide et des phosphates dans les grains d'aleurone. I. Grains d'aleurone refermont à la fois globoides et crystalloids, *J. Microsc.,* 4, 771, 1965.
50. **Dieckert, J. W., Snowden, J. E., Jr., Moore, A. T., Heinzelman, D. C., and Altschul, A. M.,** Composition of some subcellular fractions from seeds of *Arachis hypogaea, J. Food Sci.,* 27, 321, 1962.
51. **Sharma, C. B. and Dieckert, J. W.,** Isolation and partial characterization of aleurone grains of *Arachis hypogaea* seed, *Physiol. Plant,* 33, 1, 1975.
52. **Morris, G. F. I., Thurman, D. A., and Boulter, D.,** The extraction and chemical composition of aleurone grains (protein bodies) isolated from seeds of *Vicia faba, Phytochemistry,* 9, 1707, 1970.
53. **Sobolev, A. M., Suvorou, V. I., and Buzulukova, N. P.,** Isolation of aleurone grains from seeds of several plants, *Soviet Plant Physiol.,* 24, 546, 1977.
54. **Lui, N. S. T. and Altschul, A. M.,** Isolation of globoids from cottonseed aleurone grain, *Arch. Biochem. Biophys.,* 121, 678, 1967.
55. **Reddy, N. R. and Pierson, M. D.,** Isolation and partial characterization of phytic acid-rich particles from Great Northern beans *(Phaseolus vulgaris L.), J. Food Sci.,* 52, 109, 1987.
56. **Tanaka, K., Ogawa, M., and Kasai, Z.,** The rice scutellum: studies by scanning electron microscopy and electron microprobe x-ray analysis, *Cereal Chem.,* 53, 643, 1976.
57. **O'Dell, B. L. and deBoland, A.,** Complexation of phytate with proteins and cations in corn germ and oilseed meals, *J. Agric. Food Chem.,* 24, 804, 1976.
58. **Ford, J. R., Mustakas, G. C., and Schmutz, R. D.,** Phytic acid removal from soybeans by a lipid protein concentrate process, *J. Am. Oil Chem. Soc.,* 55, 371, 1978.
59. **Reddy, N. R., Sathe, S. K., and Pierson, M. D.,** Removal of phytate from Great Northern beans *(Phaseolus vulgaris L.)* and its combined density fraction, *J. Food Sci.,* 53, 107, 1988.
60. **Yao, J. J., Wei, L. S., and Steinberg, M. P.,** Effect of maturity on chemical composition and storage stability of soybeans, *J. Am. Oil Chem. Soc.,* 60, 1245, 1983.
61. **Kadam, S. S., Kute, L. S., Lawande, K. M., and Salunkhe, D. K.,** Changes in chemical composition of winged bean *(Psophocarpus tetragonolobus* L.) during seed development, *J. Food Sci.,* 47, 2051, 1982.
62. **Welch, R. M. and House, W. A.,** Availability to rats of zinc from soybean seeds as affected by maturity of seed, source of dietary protein, and soluble phytate, *J. Nutr.,* 112, 879, 1982.
63. **Welch, R. M., House, W. A., and Allaway, W. H.,** Availability of zinc from pea seeds to rats, *J. Nutr.,* 104, 733, 1974.
64. **O'Dell, B. L., deBoland, A., and Koirtyohann, R.,** Distribution of phytate and nutritionally important elements among the morphological components of cereal grains, *J. Agric. Food Chem.,* 20, 718, 1972.
65. **Simwemba, C. G., Hoseney, R. C., Variano-Marston, E., and Zeleznak, K.,** Certain B-vitamin and phytic acid contents of pearl millets *(Pennisetum americanum* [L.] Leeke), *J. Agric. Food Chem.,* 32, 31, 1984.
66. **Resurreccion, A. P., Juliano, B. O., and Tanaka, Y.,** Nutrient content and distribution in milling fractions of rice grain, *J. Sci. Food Agric.,* 30, 475, 1979.
67. **Ferguson, I. B. and Bollard, E. G.,** The movement of calcium in germinating pea seeds, *Ann. Bot. (London),* 40, 1047, 1976.
68. **Beal, L. and Mehta, T.,** Zinc and phytate distribution in peas: influence of heat treatment, germination, pH, substrate, and phosphorus on pea phytate and phytase, *J. Food Sci.,* 50, 96, 1985.
69. **Kikunaga, S., Takahashi, M., and Huzisige, H.,** Accurate and simple measurement of phytic acid contents in cereal grains, *Plant Cell Physiol.,* 26, 1323, 1985.
70. **Lolas, G. M., Palamidis, N., and Markakis, P.,** The phytic acid, total phosphorus relationship in barley, oats, soybeans, and wheat, *Cereal Chem.,* 53, 867, 1976.
71. **Harland, B. F. and Prosky, L.,** Development of dietary fiber values for foods, *Cereal Foods World,* 24, 387, 1979.
72. **Franz, K. B., Kennedy, B. M., and Fellers, D. A.,** Relative bioavailability of zinc from selected cereals and legumes using rat growth, *J. Nutr.,* 110, 2272, 1980.
73. **Morris, E. R. and Ellis, R.,** Phytate-zinc molar ratio of breakfast cereals and bioavailability of zinc to rats, *Cereal Chem.,* 58, 363, 1981.
74. **Chang, R., Schwimmer, S., and Burr, H. K.,** Phytate: removal from whole dry beans by enzymatic hydrolysis and diffusion, *J. Food Sci.,* 42, 1098, 1977.
75. **Reddy, N. R.,** Milling and Biochemical Characteristics of Triticale, M.S. thesis, Alabama A & M University, Normal, AL, 1976.
76. **Tabekhia, M. M. and Donnelly, B. J.,** Phytic acid in durum wheat and its milled products, *Cereal Chem.,* 59, 105, 1982.
77. **deBoland, A., Garner, G. B., and O'Dell, B. L.,** Identification and properties of phytate in cereal grains and oilseed products, *J. Agric. Food Chem.,* 23, 1186, 1975.
78. **Singh, B. and Sedeh, H. G.,** Characteristics of phytase and its relationship to acid phosphatases and certain minerals in triticale, *Cereal Chem.,* 56, 267, 1979.

79. **Harland, B. F. and Oberleas, D.,** Anion-exchange method for determination of phytate in foods: collaborative study, *J. Assoc. Off. Anal. Chem.,* 69, 667, 1986.
80. **Saastamoinen, M. and Heinonen, T.,** Phytic acid content of some oat varieties and its correlation with chemical and agronomical characters, *Ann. Agric. Fenn.,* 24, 103, 1985.
81. **Singh, B. and Reddy, N. R.,** Phytic acid and mineral compositions of triticales, *J. Food Sci.,* 42, 1077, 1977.
82. **Fretzdorff, B. and Weipert, D.,** Phytic acid in cereals. I. Phytic acid and phytase in rye and rye products, *Z. Lebensm. Unters. Forsch.,* 182, 287, 1986.
83. **Cilliers, J. J. L. and Van Niekerk, P. J.,** LC determination of phytic acid in food by post column colorimetric detection, *J. Agric. Food Chem.,* 34, 680, 1986.
84. **Wheeler, E. L. and Ferrel, R. E.,** A method for phytic acid determination in wheat and wheat fractions, *Cereal Chem.,* 48, 312, 1971.
85. **Doherty, C., Faubion, J. M., and Rooney, L. W.,** Semiautomated determination of phytate in sorghum and sorghum products, *Cereal Chem.,* 59, 373, 1982.
86. **Radhakrishnan, M. R. and Sivaprasad, J.,** Tannin content of sorghum varieties and their role in iron availability, *J. Agric. Food Chem.,* 28, 55, 1980.
87. **Lorenz, K.,** Tannins and phytate content in proso millets *(Panicum miliaceum), Cereal Chem.,* 60, 424, 1983.
88. **Reddy, N. R. and Salunkhe, D. K.,** Effects of fermentation on phytate phosphorus and minerals of black gram, rice, and black gram and rice blends, *J. Food Sci.,* 45, 1708, 1980.
89. **Toma, R. B. and Tabekhia, M. M.,** Changes in mineral elements and phytic acid contents during cooking of three California rice varieties, *J. Food Sci.,* 44, 619, 1979.
90. **Graf, E. and Dintzis, F. R.,** Determination of phytic acid in foods by high performance liquid chromatography, *J. Agric. Food Chem.,* 30, 1094, 1982.
91. **Vinh, L. T. and Dworschak, E.,** Phytate content of some foods from plant origin from Vietnam and Hungary, *Die Nahrung.,* 29, 161, 1985.
92. **Swaranjeet, K., Mohinder, K., and Bains, G. S.,** Chapaties with leavening and supplements: changes in texture, residual sugars and phytic acid phosphorus, *Cereal Chem.,* 59, 367, 1982.
93. **Latta, M. and Eskin, N. A. M.,** A simple and rapid colorimetric method for phytate determination, *J. Agric. Food Chem.,* 28, 1313, 1980.
94. **Ellis, R. and Morris, E. R.,** Comparison of ion-exchange and iron-precipitation methods for analysis of phytate, *Cereal Chem.,* 59, 232, 1982.
95. **Ranhotra, G. S.,** Hydrolysis during breadmaking of phytic acid in wheat protein concentrate, *J. Food Sci.,* 37, 12, 1972.
96. **Ranhotra, G. S., Loewe, R. J., and Puyat, L. V.,** Phytic acid in soy and its hydrolysis during breadmaking, *J. Food Sci.,* 39, 1023, 1974.
97. **Wallace, G. W. and Satterlee, L. D.,** Calcium binding and its effects on the properties of several food protein sources, *J. Food Sci.,* 42, 473, 1977.
98. **Nelson, K. J. and Potter, N. N.,** Iron binding by wheat gluten, soy isolate, zein, albumin, and casein, *J. Food Sci.,* 44, 104, 1979.
99. **Yoon, J. H., Thompson, L. U., and Jenkins, D. J. A.,** The effect of phytic acid on *in vitro* rate of starch digestibility and blood glucose response, *Am. J. Clin. Nutr.,* 38, 835, 1983.
100. **Harland, B. F. and Harland, J.,** Fermentative reduction of phytate in rye, white, and whole wheat breads, *Cereal Chem.,* 57, 226, 1980.
101. **Tangkongchitr, U., Seib, P. A., and Hoseney, R. C.,** Phytic acid. II. Its fate during breadmaking, *Cereal Chem.,* 58, 229, 1981.
102. **Davies, N. T. and Warrington, S.,** The phytic acid, mineral, trace element, protein, and moisture content of UK Asian immigrant foods, *Human Nutr. Appl. Nutr.,* 40A, 49, 1986.
103. **Khan, N., Zaman, R., and Elahi, M.,** Effect of processing on the phytic acid content of wheat products, *J. Agric. Food Chem.,* 34, 1010, 1986.
104. **Ceruti, G., Finoli, C., and Vecchio, A.,** Phytic acid in bran and in natural foods, *Boll. Chim. Farm.,* 123, 408, 1984.
105. **Camire, A. L. and Clydesdale, F. M.,** Analysis of phytic acid in foods by HPLC, *J. Food Sci.,* 47, 575, 1982.
106. **Davis, K.,** Proximate composition, phytic acid, and total phosphorus of selected breakfast cereals, *Cereal Chem.,* 58, 347, 1981.
107. **Reinhold, J. G.,** Phytate destruction by yeast fermentation in whole wheat meals, *J. Am. Diet. Assoc.,* 66, 38, 1975.
108. **Reinhold, J. G.,** Zinc and mineral deficiencies in man: the phytate hypothesis, in *Review of Basic Knowledge,* Chavez, A., Bourges, H., and Basta, S., Eds., S. Karger, New York, 1975, 115.
109. **Kotaru, M., Ikeuchi, T., Yoshikawa, H., and Ibuki, F.,** Investigations of antinutritional factors of the winged bean *(Psophocarpus tetragonolobus), Food Chem.,* 24, 279, 1987.
110. **Kantha, S. S. and Erdman, J. W., Jr.,** The winged bean as an oil and protein source: a review, *J. Am. Oil Chem. Soc.,* 61, 515, 1984.

111. **Kantha, S. S., Hettiarachchy, N. S., and Erdman, J. W., Jr.,** Nutrient, antinutrient contents and solubility profiles of nitrogen, phytic acid and selected minerals in winged bean flour, *Cereal Chem.,* 63, 9, 1986.
112. **Tan, N. H., Rahim, Z. H. A., Khor, H. T., and Wong, K. C.,** Winged bean *(Psophocarpus tetragonolobus* L.) tannin level, phytate content, and hemagglutinating activity, *J. Agric. Food Chem.,* 31, 916, 1983.
113. **Sutardi and Buckle, K. A.,** Phytic acid changes in soybeans fermented by traditional inoculum and six strains of *Rhizopus oligosporus, J. Appl. Bacteriol.,* 58, 539, 1985.
114. **Ologhobo, A. D. and Fetugo, B. L.,** Distribution of phosphorus and phytate in some Nigerian varieties of legumes and some effects of processing, *J. Food Sci.,* 49, 199, 1984.
115. **Gad, S. S., Mohamed, M. S., El-Zalaki, M. E., and Mohasseb, S. Z.,** Effect of processing on phosphorus and phytic acid contents of some Egyptian varieties of legumes, *Food Chem.,* 8, 11, 1982.
116. **Ferrando, R.,** Natural antinutritional factors present in European plant proteins, *Qual. Plant. Plant Foods Hum. Nutr.,* 32, 455, 1983.
117. **Eskin, N. A. M. and Wiebe, S.,** Changes in phytase activity and phytate during germination of two faba bean cultivars, *J. Food Sci.,* 48, 270, 1983.
118. **Henderson, H. M. and Ankrah, S. A.,** The relationship of endogenous phytase, phytic acid, and moisture uptake with cooking time in *Vicia faba* minro cv. Aladin, *Food Chem.,* 17, 1, 1985.
119. **Griffiths, D. W.,** The phytate content and iron-binding capacity of various field bean (*Vicia faba* L.) preparations and extracts, *J. Sci. Food Agric.,* 33, 847, 1982.
120. **Kon, S. and Sanshuck, D. W.,** Phytate content and its effect on cooking quality of beans, *J. Food Process. Preserv.,* 5, 169, 1981.
121. **Chen, L. H. and Pan, S. H.,** Decrease of phytates during germination of pea seeds *(Pisum sativa), Nutr. Rep. Intern.,* 46, 125, 1977.
122. **Sharma, Y. K., Tiwari, A. S., Rao, K. C., and Mishra, A.,** Studies on chemical constituents and their influences on cookability in pigeon pea, *J. Food Sci. Technol., (India)* 14, 38, 1977.
123. **Rao, P. U. and Deosthale, Y. G.,** Effect of germination and cooking on mineral composition of pulses, *J. Food Sci. Technol. (India),* 20, 195, 1983.
124. **Singh, U., Kherdekar, M. S., Sharma, D., and Saxena, K. B.,** Cooking quality and chemical composition of some early, medium, and late maturing cultivars of pigeon pea (*Cajanus cajan* L. Mill.), *J. Food Sci. Technol. (India),* 21, 367, 1984.
125. **Kumar, K. G., Venkataraman, L. V., Jaya, T. V., and Krishnamurthy, K. S.,** Cooking characteristics of some germinated legumes: Changes in phytins, Ca^{++}, Mg^{++}, and pectins, *J. Food Sci.,* 43, 85, 1978.
126. **Ologhobo, A. D. and Fetuga, B. L.,** Investigations on the trypsin inhibitor, hemagglutinin, phytic and tannic acid contents of cowpea (*Vigna unguiculata* L.), *Food Chem.,* 12, 249, 1983.
127. **Elkowicz, K. and Sosulski, F.,** Antinutritive factors in eleven legumes and their air classified protein and starch fractions, *J. Food Sci.,* 47, 1301, 1982.
128. **Khokhar, S. and Chauhan, B. M.,** Antinutritional factors in moth bean *(Vigna aconitifolia)*: varietal differences and effects of methods of domestic processing and cooking, *J. Food Sci.,* 51, 59, 1986.
129. **Reddy, N. R., Balakrishnan, C. V., and Salunkhe, D. K.,** Phytate phosphorus and mineral changes during germination and cooking of black gram (*Phaseolus mungo* L.) seeds, *J. Food Sci.,* 43, 540, 1978.
130. **Ologhobo, A. D. and Fetuga, B. L.,** Polyphenols, phytic acid and other phosphorus compounds of lima beans (*Phaseolus lunatus* L.), *Nutr. Rep. Intern.,* 26, 605, 1982.
131. **Knuckles, B. E., Kuzmicky, D. D., and Betschart, A. A.,** HPLC analysis of phytic acid in selected foods and biological samples, *J. Food Sci.,* 47, 1257, 1982.
132. **Lolas, G. M. and Markakis, P.,** Phytic acid and other phosphorus compounds of beans (*Phaseolus vulgaris* L.), *J. Agric. Food Chem.,* 23, 13, 1975.
133. **Iyer, V. G., Salunkhe, D. K., Sathe, S. K., and Rockland, L. B.,** Quick-cooking beans (*Phaseolus vulgaris* L.). II. Phytates, oligosaccharides, and antienzymes, *Qual. Plant. Plant Foods Hum. Nutr.,* 30, 45, 1980.
134. **Deshpande, S. S. and Cheryan, M.,** Changes in phytic acid, tannins, and trypsin inhibitory activity on soaking of dry beans (*Phaseolus vulgaris* L.), *Nutr. Rep. Intern.,* 27, 371, 1983.
135. **Lolas, G. M. and Markakis, P.,** The phytase of navy bean (*Phaseolus vulgaris* L.), *J. Food Sci.,* 42, 1094, 1977.
136. **Deshpande, S. S., Sathe, S. K., Salunkhe, D. K., and Cornforth, D. P.** Effects of dehulling on phytic acid, polyphenols, and enzyme inhibitors of dry beans (*Phaseolus vulgaris* L.), *J. Food Sci.,* 47, 1846, 1982.
137. **Sathe, S. K., Deshpande, S. S., Reddy, N. R., Goll, D. E. and Salunkhe, D. K.,** Effects of germination on proteins, raffinose oligosaccharides, and antinutritional factors in the Great Northern beans (*Phaseolus vulgaris* L.), *J. Food Sci.,* 48, 1796, 1983.
138. **Baker, E. C., Mustakas, G. C., Erdman, J. W., Jr., and Black, L. T.,** The preparation of soy products with different levels of native phytate for zinc bioavailability studies, *J. Am. Oil Chem. Soc.,* 58, 541, 1981.
139. **Pucciano, M. F., Weingartner, K. E., and Erdman, J. W., Jr.,** Relative bioavailability of dietary iron from three processed soy products, *J. Food Sci.,* 49, 1558, 1984.
140. **Thompson, D. B. and Erdman, J. W., Jr.,** Phytic acid determination in soybeans, *J. Food Sci.,* 47, 513, 1982.

141. **Harland, B. F. and Oberleas, D.,** A modified method for phytate analysis using an ion-exchange procedure: application to textured vegetable proteins, *Cereal Chem.,* 54, 827, 1977.
142. **Naczk, M., Rubin, L. J., and Shahidi, F.,** Functional properties and phytate content of pea protein preparations, *J. Food Sci.,* 51, 1235, 1986.
143. **Winters, D. D., Handel, A. P., and Lohrberg, J. D.,** Phytic acid content of crude, degummed and retail soybean oils and its effect on stability, *J. Food Sci.,* 49, 113, 1984.
144. **Madhusudhan, K. T. and Singh, N.,** Studies on linseed proteins, *J. Agric. Food Chem.,* 31, 959, 1983.
145. **Arntfield, S. D., Ismond, M. A. H., and Murray, E. D.,** The fate of antinutritional factors during the preparation of a faba bean protein isolate using a micellization technique, *Can. Inst. Food Sci. Technol. J.,* 18, 37, 1985.
146. **Anno, T., Nakanishi, K., Matsuno, R., and Kamikubo, T.,** Enzymatic elimination of phytate in soybean milk, *Nippon Shokuhin Kogyo Gakkai-Shi,* 32, 174, 1985.
147. **Sutardi and Buckle, K. A.,** Reduction in phytic acid levels in soybeans during tempeh production, storage, and drying, *J. Food Sci.,* 50, 260, 1985.
148. **Reddy, N. R.,** Effects of Cooking and Germination of Black Gram and Fermentation of Black Gram and Rice Blend on Phytates, Alpha-Amylase Inhibitor, Phytohemagglutinins, and Flatulence-Producing Factors, Ph.D. dissertation, Utah State University, Logan, 1981.
149. **Rajalakshmi, R. and Vanaja, K. V.,** Chemical and biological evaluation of effects of fermentation on the nutritive value of foods prepared from rice and grams, *Br. J. Nutr.,* 21, 467, 1967.
150. **Ramakrishnan, C. V.,** Studies on Indian fermented foods, *Baroda J. Nutr.,* 6, 1, 1979.
151. **Fardiaz, D. and Markakis, P.,** Degradation of phytic acid in oncom (fermented peanut press cake), *J. Food Sci.,* 46, 523, 1981.
152. **Reddy, N. R. and Salunkhe, D. K.,** Interactions between phytate, protein, and minerals in whey fractions of black gram, *J. Food Sci.,* 46, 564, 1981.
153. **Bassiri, A. and Nahapetian A.,** Differences in concentrations and interrelationships of phytate, phosphorus, magnesium, calcium, zinc, and iron in wheat varieties grown under dry land and irrigated conditions, *J. Agric. Food Chem.,* 25, 1118, 1977.
154. **Nahapetian, A. and Bassiri, A.,** Variations in concentrations and interrelationships of phytate, phosphorus, magnesium, calcium, zinc, and iron in wheat varieties during two years, *J. Agric. Food Chem.,* 24, 947, 1976.
155. **Miller, G. A., Youngs, V. L., and Oplinger, E. S.,** Environmental and cultivar effects on oat phytic acid concentration, *Cereal Chem.,* 57, 189, 1980.
156. **Miller, G. A., Youngs, V. L., and Oplinger, E. S.,** Effect of available soil phosphorus and environment on the phytic acid concentration in oats, *Cereal Chem.,* 57, 192, 1980.
157. **Proctor, J. P. and Watts, B. M.,** Effect of cultivar, growing location, moisture, and phytate content on the cooking times of freshly harvested navy beans, *Can. J. Plant Sci.,* 67, 923, 1987.
158. **Griffiths, D. W. and Thomas, T. A.,** Phytate and total phosphorus content of field beans (*Vicia faba* L.), *J. Sci. Food Agric.,* 32, 187, 1981.
159. **Sorensen, C. and Truelsen, E.,** Chemical composition of barley varieties with different nutrient supplies. I. Concentration of nitrogen, tannins, phytate, β-glucans, and minerals, *Tidsskr. Plantearl.,* 89, 253, 1985.
160. **Saastamoinen, M.,** Effect of nitrogen and phosphorus fertilization on the phytic acid content of oats, *Cereal Res. Commun.,* 15, 57, 1987.
161. **Harland, B. F. and Peterson, M.,** Nutritional status of lacto-ovovegetarian Trappist Monks, *J. Am. Diet. Assoc.,* 72, 259, 1978.
162. **Ellis, R., Kelsay, J. L., Reynolds, R. D., Morris, E. R., Moser, P. B., and Frazier, C. W.,** Phytate:zinc and phytate × calcium:zinc millimolar ratios in self selected diets of Americans, Asian Indians, and Nepalese, *J. Am. Diet. Assoc.,* 87, 1043, 1987.
163. **Davies, N. T.,** Effects of phytic acid on mineral availability, in *Dietary Fiber in Health and Disease,* Vahoung, G. V., and Kritchevsky, D., Eds., Plenum Press, New York, 1982, 105.
164. **Carnovale, E., Lombardi-Boccia, G., and Lugaro, E.,** Phytate and zinc content of Italian diets, *Hum. Nutr. Appl. Nutr.,* 41A, 180, 1987.
165. **Bindra, G. S., Gibson, R. S., and Thompson, L. U.,** (Phytate) × (calcium)/(zinc) ratios in Asian immigrant lacto-ovovegetarian diets and their relationship to zinc nutriture, *Nutr. Res.,* 6, 475, 1986.
166. **Mbofung, C., Atinmo, T., and Omololu, A.,** Zinc and phytate concentrations, phytate:zinc molar ratio, and metallo calorie ratio of zinc and protein contents of some selected Nigerian dietary foods, *Nutr. Res.,* 4, 567, 1984.
167. **Harland, B. F. and Felix-Phipps, R.,** The implications of high phytate in Nigerian foods, *Fed. Proc.,* 46, 1003 (abstr.), 1987.
168. **Reddy, N. R., Sathe, S. K., and Salunkhe, D. K.,** Phytates in legumes and cereals, *Adv. Food Res.,* 28, 1, 1982.
169. **Ellis, R. and Morris, E. R.,** Improved ion-exchange phytate method, *Cereal Chem.,* 60, 120, 1983.
170. **O'Neill, I. K., Sargent, M., and Trimble, M. L.,** Determination of phytate in foods by phosphorus[31] Fourier Transform Nuclear Magnetic Resonance Spectrometry, *Anal. Chem.,* 52, 1288, 1980.

171. **Davis, K.,** Effect of processing on composition and *Tetrahymena* relative nutritive value of green and yellow peas, lentils, and white pea beans, *Cereal Chem.,* 58, 454, 1981.
172. **Srivastava, B. N., Biswas, T. D., and Das, N. B.,** Influence of fertilizers and manures on the content of phytin and other forms of phosphorus in wheat and their relation to soil phosphorus, *J. Ind. Soc. Soil Sci.,* 3, 33, 1955.
173. **Raboy, V. and Dickinson, D. B.,** Effect of phosphorus and zinc nutrition on soybean seed phytic acid and zinc, *Plant Physiol.,* 75, 1094, 1984.
174. **Batten, G. D. and Lott, J. N. A.,** The influence of phosphorus nutrition on the appearance and composition of globoid crystals in wheat aleurone cells, *Cereal Chem.,* 63, 14, 1986.
175. **Michael, B., Zin, F., and Lantzsch, H.,** Effect of phosphate application on phytin-phosphorus and other phosphate fractions in developing wheat grains, *Z. Pflanzenernaehr. Bodenkd.,* 143, 369, 1980.
176. **Raboy, V. and Dickinson, D. B.,** The timing and rate of phytic acid accumulation in developing soybean seeds, *Plant Physiol.,* 85, 841, 1987.
177. **Frolich, W. and Nyman, M.,** Minerals, phytate, and dietary fibre in different fractions of oat grain, *J. Cereal Sci.,* 7, 73, 1988.
178. **Kirby, L. K. and Nelson, T. S.,** Total and phytate phosphorus content of some feed ingredients derived from grains, *Nutr. Rep. Intern.,* 137, 277, 1988.
179. **Harland, B. F. and Oberleas, D.,** Phytate in foods, *World Rev. Nutr. Diet.,* 52, 235, 1987.
180. **Carnovale, E., Lugaro, E., and Lombardi-Boccia, G.,** Phytic acid in faba bean and pea: effect on protein availability, *Cereal Chem.,* 675, 114, 1988.
181. **Murphy, S. P. and Calloway, D. H.,** Nutrient intakes of women in NHANES II, emphasizing trace minerals, fiber, and phytate, *J. Am. Diet. Assoc.,* 86, 1366, 1986.
182. **Torelm, I. and Bruce, A.,** Phytic acid in foods, *Var Foda,* 34, 79, 1982.
183. **Wise, A., Lockie, G. M., and Liddell, J.,** Dietary intakes of phytate and its meal distribution pattern amongst staff and students in an institution of higher education, *Br. J. Nutr.,* 58, 337, 1987.
184. **Rao, P. U. and Deosthale, Y. G.,** *In vitro* availability of iron and zinc in white and colored ragi *(Eleusine coracana):* role of tannin and phytate, *Plant Foods Hum. Nutr.,* 38, 35, 1988.
185. **Kadam, S. S., Smithard, R. R., Eyre, M. D., and Armstrong, D. G.,** Effect of heat treatments on antinutritional factors and quality of proteins in winged bean, *J. Sci. Food Agric.,* 39, 267, 1987.

Chapter 7

INTERACTIONS OF PHYTATE WITH PROTEINS AND MINERALS

I. INTRODUCTION

Phytic acid in the free acid form is unstable and decomposes to yield orthophosphoric acid[1] but it is quite stable when it is present in dry form as its salt. There are twelve replaceable hydrogen atoms on phytic acid.[2] Six of these twelve are strongly dissociable at pKa of about 1.84, two weakly dissociable at pKa of 6.30, and the remaining four being so weakly dissociable at pKa of 9.70 that they could not be titrated. Costello et al.,[3] using ^{31}P nuclear magnetic resonance (^{31}PNMR) -pH titration methods, came to similar conclusions (6 protons pKa = 1.1 to 2.1, one proton pKa = 5.70, two protons pKa = 6.80 to 7.60, and three protons with pKa = 10.0 to 12.0). Recently, Evans et al.[4] found six protons with pKa = 2.18, two with pKa = 5.73, and four with pKa = 9.21 by titrating phytic acid or potassium phytate in the presence of 0.2M KCl and concluded that the presence/absence of such an electrolyte in titration studies can influence the pKa values. These studies suggest that the phytic acid has a potential for binding positively charged proteins and/or multivalent cations or minerals in foods, since it occurs as a strongly negatively charged molecule over a wide range of pH values. The resultant complexes formed between phytate, protein, and minerals in foods may be nutritionally unavailable for absorption under normal physiological conditions. Thus, a concern arises about its presence in cereals, legumes, and their processed products. The interactions between phytate, protein, and minerals that occur during various processing conditions are discussed in this chapter. The influence of phytate on the functional and nutritional properties of proteins is also covered.

II. PHYTATE-PROTEIN INTERACTIONS

The subject of phytate-protein interaction(s) has been studied for over forty years to answer questions including: (1) how phytate influences the protein solubility in aqueous media, (2) the forces and sites which are responsible for phytate-protein interaction(s), (3) how the environmental manipulations affect these interactions, (4) how to remove protein-bound phytate to facilitate the preparation of phytate-free protein, and (5) how phytate inhibits enzymes and the nutritional significance of such enzyme inhibition in animal and human nutrition.

A. FORMATION OF PHYTATE-PROTEIN COMPLEXES

Phytic acid interacts with proteins at various pH values to form phytate-protein complexes.[5] These interactions can be grouped into three categories[5] based on pH: (1) low pH, (2) intermediate pH, and (3) high pH.

1. Low pH (3.5 or Less)

At a low acidic pH, phytic acid has a strong negative charge, while many plant proteins would be positively charged since their isoelectric pH is near pH 4.0 to 5.0. This makes it possible for these compounds to interact with each other; the $-NH_3^+$ groups on protein are bound to the phosphate groups of phytic acid (Figure 1). The nature of the participating bonds, although not yet proven, could be either ionic, hydrogen or van der Waal type. The $-NH_3^+$ groups on proteins would include the amino terminal (if free), ε-amino group of lysine, imidazole group of histidine, and guanidyl group of arginine.[5] Divalent metal ions such as Ca^{2+} may also interact with phytic acid under these acidic pH conditions. The metal binding may be between two phosphate groups of phytic acid.

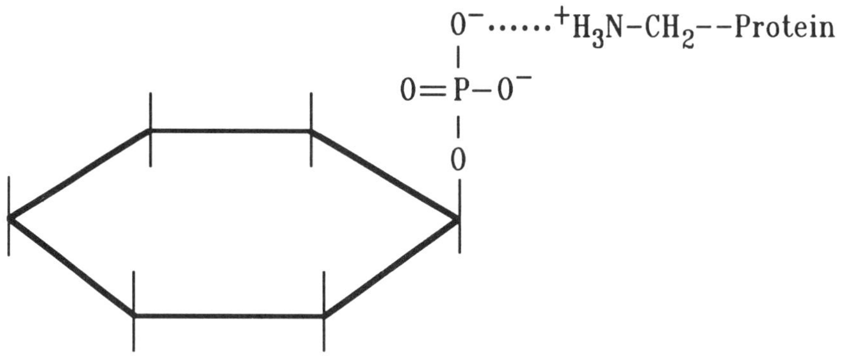

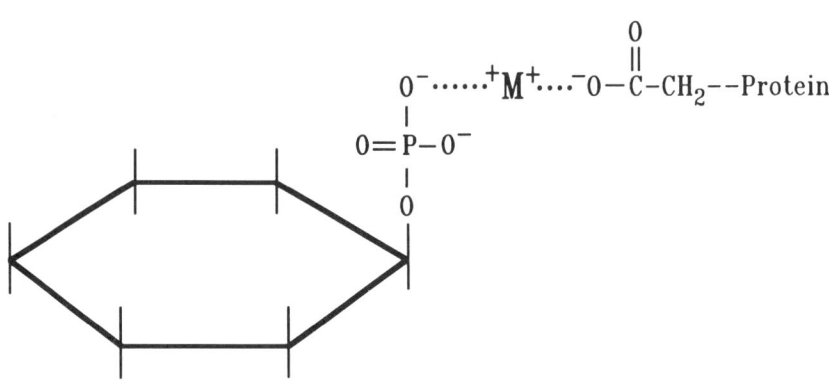

FIGURE 1. Possible structure of phytic acid-protein complex at low pH. (From Cheryan, M., *CRC Crit. Rev. Food Sci. Nutr.*, 13, 297, 1980. With permission.)

FIGURE 2. Possible structure of phytic acid-protein complex at alkaline pH. (From Cheryan, M., *CRC Crit. Rev. Food Sci. Nutr.*, 13, 297, 1980. With permission.)

2. Intermediate pH

In the intermediate pH range both phytic acid and proteins have a net negative charge, provided the isoelectric pH of the protein under consideration is lower than the pH being considered. Under these conditions, interaction between two negatively charged protein and phytic acid molecules would not theoretically be possible. However, complexation occurs between phytic acid and proteins under these conditions. Possible mechanisms include (1) a direct binding of phytic acid to α-NH_2 terminal group and ε-NH_2 groups of lysine because they are still protonated and (2) a multivalent cation-mediated interaction[5,6] (Figure 2).

$$\text{cation} + \text{phytic acid} \Leftrightarrow (\text{cation-phytic acid})$$

$$\text{protein} + \text{cation} + \text{phytic acid} \Leftrightarrow (\text{protein-cation-phytic acid})$$

Cation binding to protein molecules could be via binding to carboxyl or histidyl groups. Some binding may also occur without the mediation by metal ions because the lysyl and arginyl side chains may remain protonated in the intermediate pH range.[7] Direct binding between phytic acid and basic proteins would also be possible under such conditions.

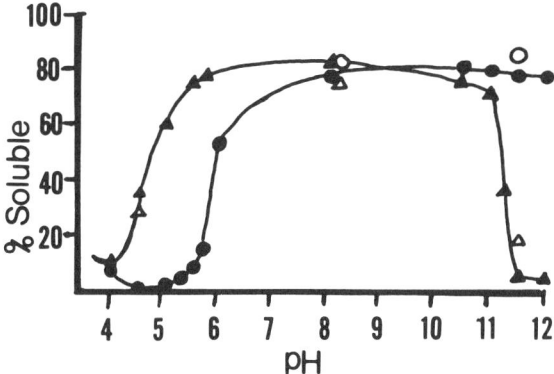

FIGURE 3. Influence of pH on the solubility of nitrogen and phytate of soy flour, nitrogen (○ -- ○), and phytate (▲ -- ▲). Closed symbols indicate laboratory trials while open symbols denote pilot plant trials. (From de Rham, O. and Jost, T., *J. Food Sci.*, 44, 596, 1979. With permission.)

3. High pH

At high pH values (pH >9.00), the nature of interaction between phytic acid and protein is not clearly understood.[5] There are indications that when the pH is high, the interaction between phytic acid and protein is diminished.

B. INTERACTIONS OF PHYTATE WITH SOY PROTEINS

Early studies of Fontaine et al.[8] indicated that phytates can interact with soybean proteins under certain conditions and that these interactions are affected by characteristic properties of both the protein(s) and phytic acid. The major conclusions of this study were: (1) phytic acid decreased the solubility of seed proteins at pH values below their isoelectric point, (2) in isolated seed proteins, phytic acid was the major impurity, the amount being dependent on the extraction pH, (3) phytate may influence the clarity of certain seed meal extracts. Further, Fontaine et al.[8] showed that about 40 to 45% of the total phosphorus was solubilized at the isoelectric point of soy, cottonseed, and peanut proteins. Smith and Rackis[9] illustrated the influence of phytate-protein interactions on the properties of soy proteins. They demonstrated that the removal of phytin from soybean extracts caused an increase of 0.8 unit in the isoelectric point of soy proteins. Subsequently, several investigators[10-12] confirmed the finding that there is a decrease in solubility of soy proteins in the presence of phytate. A typical solubility curve for soy flour in the presence of phytate is shown in Figure 3.

Saio et al.[10] systematically evaluated the interactions between phytate and soy proteins in the presence and absence of calcium. The reason for including calcium in their studies was the fact that the amount of phosphorus associated with the cold insoluble fraction of soy proteins (11S or glycinin) was significantly different (1.40% phosphorus compared to 0.17% phosphorus precipitated in the presence and absence of calcium respectively).[13] This suggests that the 11S protein in soybean could bind more phytic acid phosphorus in the presence of calcium. Subsequently, studies by Saio et al.[14,15] confirmed that in the presence of calcium, phytic acid could bind to soy proteins even at pH 6.6, the pH at which soybean phosphorus is readily soluble. Okubo et al.[16] further investigated the interactions between glycinin of soybeans and phytic acid using gel filtration (Sephadex® G-75), turbidimetric titrations, and manipulation of environment (pH and presence or absence of calcium). They did not detect binding between the phytic acid and glycinin, above the isoelectric point of glycinin at pH 6, 8, and 10. However, several important findings emerged from the work of Okubo et al.[16] These include: (1) insoluble complex formation between glycinin and phytic acid occurred between pH 2.5 and 5.0. The extent of binding increased from 0 at the isoelectric point to a maximum of 424 equivalents of phytate molecules per mole of glycinin (11S dimer with molecular weight 360,000 daltons) as the pH was decreased. The equivalents of phytate binding were calculated on the basis of 411 cationic groups (12 free amino terminal groups plus the number of basic amino acid residues) of glycinin dimer; (2) the quarternary structure of glycinin did influence the binding of phytate molecules to it (accessibility of charged groups changed with the change in structure of glycinin); and (3) calcium ions promoted the dissociation of phytate-glycinin complexes at pH 3.0, presumably as a result of competition between calcium and protein cationic groups for the phosphate groups of phytate molecule. A complete dissociation of phytate-glycinin complex requires a 105-fold equivalent excess of calcium with respect to the cationic groups of glycinin. O'Dell and deBoland,[17] however, could not satisfactorily correlate basic amino acid content of corn germ and sesame seed proteins and phytate binding. However, they did find phytate could bind to soy proteins readily. It is, therefore, important to note that the binding of phytate molecules to proteins may depend upon additional factors such as the naturally occurring form of phytate in a given source, the type and solubility of the protein(s) under consideration, and the conformational characteristics of the protein(s) molecule(s).

The observation of Okubo et al.[16] that in the absence of calcium a very small amount of phytate can bind to glycinin has been further supported by Brooks and Morr.[18] These later researchers fractionated soy proteins into 11S, 7S, soy whey precipitate, and soy whey supernatant. Although they did not purify the 11S, their data on crude 11S preparation showed that the 11S fraction had very little bound phytate (phytate to protein ratio was 0.0007 and phytate to phosphorus ratio was 0.875). These data clearly indicate that a divalent metal ion such as calcium must be facilitating the phytate binding to 11S protein of soybeans. Appurao and Narasingarao[19] studied the interactions between calcium and the glycinin of soybeans by equilibrium dialysis at pH 5.5 and 7.8 and concluded that the binding was negligible at pH 5.5 and significant at pH 7.8. The binding of calcium at pH 7.8 was significantly decreased when $0.5M$ NaCl was added. They also noted that when calcium was added to unfractionated soy proteins, it bound to proteins as well as the phytate associated with proteins. When they compared the calcium binding to unfractionated soy proteins (containing phytate) vs. the phytate-free proteins (in $0.1M$ borate buffer, pH 7.8), the phytate-free proteins bound much less calcium, clearly indicating the role of calcium in the phytate-metal-protein complexation. Wallace and Satterlee[20] similarly explained calcium binding to soy protein isolate. One possible mechanism by which such complexation may be mediated is the alkaline-earth bridge formation, where the divalent metal ion is bridged between the phosphate group of phytate and the carboxyl group of the protein (Figure 2).

Relatively few data appear in the literature on the phytate binding to other storage proteins (namely 7S, β-conglycinin). Brooks and Morr[18] have recently shown that in soy proteins, the 7S and whey fractions (fractionation was based on differential solubility of soy proteins in a pH-dependent manner) accounted for the major amount of soybean phytate with negligible binding to the 11S fraction (Table 1). Although their 11S and 7S fractions were not chromatographically purified and consequently must have 11S in the 7S fraction, and the fact that 11S does not bind phytate to a significant extent clearly points toward the importance of 7S and whey proteins in phytate binding. Prattley and Stanley[21] chromatographed the soybean proteins at pH 6.8 ($0.05M$ Tris buffer) and found that calcium and phytate were associated with the 7S peak. In contrast, Honig and Wolf[22] reported that most minerals and phytic acid did not elute as a complex with the soybean 7S protein. This apparent discrepancy can be due to the use of two different chromatography columns, material, and conditions. A closer examination of the chromatography profile of Prattley and Stanley[21] does indicate that the phosphorus peak eluted slightly after the 7S protein peak. The association of phytate with soy whey in higher amounts (2.066%)

TABLE 1
Phytate Content of Soy Protein Fractions[18]

Protein fraction	Apparent protein content (%)	Phosphorus content (%)	Phytate content (%)	Phytate to protein ratio	Phytate to phosphorus ratio
Commercial defatted soy flakes whole extract	78.95	0.60	1.41	0.0179	2.35
7S fraction	112.60	0.68	1.40	0.0124	2.05
11S fraction	100.80	0.08	0.07	0.0007	0.88
Soy whey	16.88	0.25	0.28	0.0166	1.12
Soy whey precipitate	4.49	15.20	45.37	10.1046	2.99

compared to soy curd (1.024%) when expressed on a moisture-free basis[23] is consistent with the earlier observations of Brooks and Morr.[18]

The foregoing discussion suggests that under neutral to slightly alkaline conditions, the major amount of phytate in soybeans is associated with whey, to a lesser extent with 7S protein fraction, and negligible amounts with the 11S protein fraction.

C. INTERACTIONS OF PHYTATE WITH OTHER PROTEINS

O'Dell and deBoland[17] studied the importance of naturally occurring chemical forms of phytic acid with respect to its reactivity towards protein and minerals in corn germ. About 90% of phytate in corn germ is water soluble while only 40% of corn germ proteins are water soluble. The data of O'Dell and deBoland[17] indicated that when the corn germ proteins were isoelectrically precipitated (pH 4.8), only 20% of the corn germ phytate was coprecipitated with the proteins. Phytate was not shown to be bound to the proteins to any significant extent because they could remove the phytate from the isoelectrically precipitated protein fraction by electrophoresis (discontinuous electrophoresis in 7.5% acrylamide gel in tris-glycine buffer system of pH 8.9, running buffer pH of 9.3). Further, they[17] concluded that corn germ albumins (which are rich in lysine and arginine) do not interact with phytate. To further establish that under acidic conditions there was no interaction between phytate and protein, these investigators[17] chromatographed water soluble corn germ proteins from a high lysine corn germ on a Sephadex G-50 column in an acetate buffer system (pH 4.40) and found that phytate eluted at the column volume just as pure phytate would in the absence of corn germ proteins.

Coprecipitation of phytate with peanut and cotton seed proteins, when proteins were precipitated at their isoelectric points, has also been reported.[8] The solubilities of phosphorus, protein, and inorganic phosphorus for defatted peanuts, cottonseeds, and soybeans are presented in Figure 4. They noted that similar to soybean proteins, phosphorus solubility paralleled that of nitrogen solubility, especially in the pH range of 1.5 to 3.5, indicating that most of the phosphorus (85%) in peanuts is associated with proteins at pH 3.50 (below isoelectric point of peanut proteins). As the pH was increased (from 3.5 to 6.5), total phosphorus solubility increased very rapidly compared to that of nitrogen, i.e., at pH 5.80, only 13% nitrogen was solubilized, compared to 86.6% for phosphorus solubility. The phosphorus associated with isoelectrically precipitated peanut proteins could not be washed/dialyzed away and, therefore, these researchers[8] concluded that phytate must be complexing with the peanut proteins below the isoelectric pH of proteins. The solubility pattern of cottonseed proteins is distinctly different than those for peanut and soybean. This illustrates further that the properties of phytate are different (in cottonseeds, insoluble salts at pH 2.75 with maximum solubility for total phosphorus at pH 0.7 and 6.0) depending on the source. Hellot and Macheboeuf[24] had similarly indicated the association of phytate with peanut proteins, when the proteins were precipitated by pH adjustment of the extract.

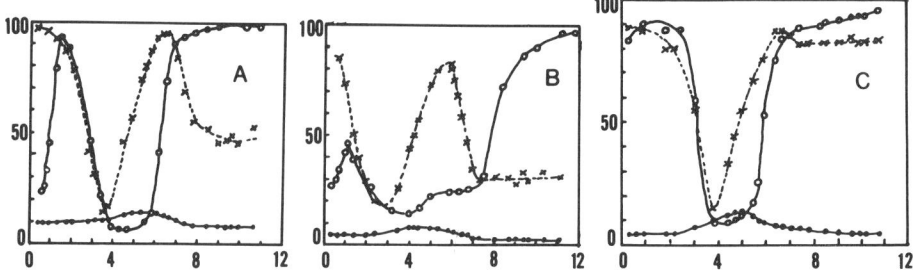

FIGURE 4. Solubilities of defatted peanut (A), cottonseed (B), and soybean (C) nitrogen (○ - - ○), total phosphorus (× - - ×), and inorganic phosphorus (● - - ●). All data are expressed as % (y-axis) as a function of pH (x-axis). The inorganic phosphorus values are a % of total meal phosphorus. (From Fontaine, T.D., Pons, W. A., and Irving, G. W., *J. Biol. Chem.* 164, 487, 1946. With permission.)

In wheat flour proteins, very little association between phytate and proteins has been found.[25,26] In a more detailed study on rice bran, Champagne et al.[27,28] studied the interaction between phytate, minerals, and rice bran proteins. Rice bran phytate cannot form complexes with rice bran proteins above pH of 2.0. They[27,28] concluded that the solubility profiles of phosphorus and nitrogen as a function of pH did not correspond except in a pH range of 1.0 to 2.0, an observation similar to the solubility of phosphorus and nitrogen in wheat bran proteins reported earlier.[25] The solubility of potassium, magnesium, and calcium corresponds to the solubility of phytic acid as a function of pH, while those of iron, zinc, and copper had no such correlation.

Bourdillon[29] prepared a crystalline phytate-protein complex formed at pH 4.0 from Great Northern beans. Recently, Reddy and Pierson[30] indicated that in Great Northern beans, phytate occurs as water-soluble as well as water-insoluble salts in association with proteins, indicating possible complexation of phytate and protein *in vivo*.

Reddy and Salunkhe[31] studied the interactions between phytate, minerals, and whey proteins in black gram at pH 2.80, 6.40, and 8.40. This study was based on dialysis of black gram whey (water-extracted black gram proteins were precipitated at pH 3.50 and the resulting whey was dialyzed at pH 2.80, 6.40, and 8.40 to obtain fractions I, II, and III, respectively) fractions for 2 days. Maximum complexation between phytate and whey proteins occurred at pH 8.40 (fraction II) and least at pH 6.40 (Table 2).

The binding constants for phytic acid with human serum albumin and ovalbumin were 10^6 and 5×10^4, respectively.[32] Prattley et al.[33] obtained a binding constant of 2.3×10^5 for phytic acid and bovine serum albumin at pH 3.0. Barré and Van Huot[32] found that an agreement between the number of binding sites (the basic amino acids, lysine, arginine, and histidine) on the proteins and the total number of phytic acid molecules bound to those proteins (per mole of protein). In contrast, Prattley et al.[33] did not find such an agreement. They[33] observed that of the 93 potential binding sites in the form of basic amino acids in the bovine serum albumin (lysine 58, arginine 19, and histidine 16), only 78 sites were occupied by phytic acid at pH 3.0. Further, they postulated that the difference (15 potential binding sites) must be due to occlusion of these sites within the molecule, thus making them unavailable for phytic acid binding. This suggests that the protein conformation must have some role in controlling the phytic acid-protein interactions.

D. EFFECTS OF PHYTATE-PROTEIN INTERACTIONS ON PROTEIN FUNCTIONALITY

From the foregoing discussion, it is evident that the interaction of phytate with proteins may have certain effects on functional properties of proteins including protein solubility, charge, and possibly structure.[7] The main functional property affected by phytate is the solubility.[5] The

TABLE 2
Complexation of Phytate with Proteins in Fractions I, II, and III as Measured by Dialysis at Different pH and Media[31]

Sample	pH	Medium	Day of dialysis 0	Day of dialysis 2	Ratio mg phytate/ g protein at 2 days
			% Phytate retained		
Fraction I	2.80	HCl (0.002 M)	100[a]	45 (21.5%)[b]	164.2
Fraction II	8.40	Tris buffer (0.05 M)	100	69 (33.2%)	144.0
Fraction III	6.40	Distilled water	100	4 (1.9%)	28.4

[a] Whey fraction, undialyzed (pH 3.50) had a ratio of 201.6 mg phytate/g protein.
[b] Values in parenthesis are the percentages recovered from original bean meal.

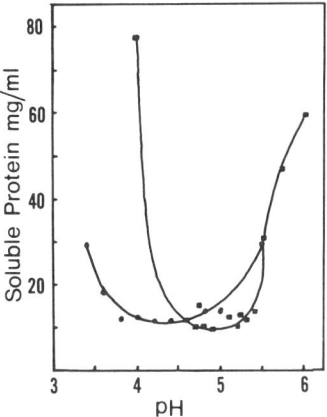

FIGURE 5. Protein solubility of soy extracts as influenced by the pH. Laboratory control extract (● - - ●), phytate reduced extract (■ - - ■) (From Chen, B. H. Y. and Morr, C. V., *J. Food Sci.*, 50, 1139, 1985. With permission.)

reduction in solubility of proteins caused by the presence of phytate has been discussed in the earlier part of this chapter. The interaction of phytate with proteins results in a change in the isoelectric point of the protein(s). On the other hand, removal of phytate reduces the tendency of protein to aggregate in water dispersions, increases the isoelectric point of the protein, and improves the solubility of the protein on the acid side of the isoelectric point.[7,34] Chen and Morr[12] have shown that when soy protein isolate was treated with ion-exchange resin to reduce phytate content, the resulting protein isolate (with phytate content of 0.43%) had a minimum protein solubility at a higher pH compared to the isolate which was not pretreated for phytate removal (phytate content of 1.90%) (Figure 5). The solubility of protein isolate containing low amounts of phytate was higher at pH 4.0 compared to the control, which had higher phytate content. They[12] suggest that for the low-phytate isolate the improved solubility of soy proteins below the isoelectric pH was due to unmasking of positive charges providing additional sites for protein hydration. Further, Chen and Morr[12] evaluated foaming capacity and stability of the soy isolates and found that the phytate-reduced soy protein isolates had a larger mean optimum foam expansion value of 1336% compared to 984% for the control isolate. The foaming performance

TABLE 3
Functional Properties of Gluten, Soy, and Pea Protein Products[35]

Protein product	Phytate (%)	Water absorption[a] (%)	Water hydration capacity (g H$_2$O/g)	Fat absorption[b] (%)	Protein[c] solubilization (%)	Emulsifying[d] activity (%)
Gluten protein preparation Wheatpro-80® (Code 6006)	0.27	152	1.37	96.3	16.3	66.5
Soy protein concentrate, PROMAX 70®	1.17	445	5.52	157.0	31.5	59.4
Soy protein isolate, SUPRO 620®	1.69	584	5.85	144.0	22.2	75.1
Woodstone pea protein preparation (batch #2328)	2.25	287	3.21	94.5	30.3	71.5
Woodstone pea protein preparation (batch #2521)	2.00	278	2.80	93.7	37.7	70.7
Woodstone pea protein preparation (batch # Unknown)	2.00	293	3.13	90.1	41.9	68.5

[a] Used aqueous dispersions containing 2 g protein product in 16 ml distilled water, contents mixed every 10 min for 30 sec. (total of 7 mixings), and centrifuged at 2000 × g for 15 min.
[b] A 2 g sample was dispersed in 12 ml of soybean oil in a 25 ml centrifuge tube to determine fat absorption by centrifugation method.
[c] A 2.5 g sample in 200 ml distilled water was shaken for 2 h at 22°C, 100 ml of this was centrifuged at 365 ×g (1500 rpm), supernatant filtered through glass wool, and soluble protein on the filtrate determined.
[d] A 3.5 g sample was homogenized for 30 sec. with 50 ml distilled water, 25 ml soybean oil added, homogenized again for 90 sec, and then centrifuged at 1100 × g for 5 min after dividing the sample into four even parts in 25 ml centrifuge tubes. Emulsifying activity was calculated as the percentage ratio of the volume of the emulsified layer to that of emulsion prior to centrifugation.

of the control isolate was best at pH 6.0, whereas that of the phytate-reduced isolate was at pH 3.0.

Naczk et al.[35] compared the functional properties of several pea protein preparations with those of wheat gluten, soy isolate, and a soy concentrate. The pea protein preparations had a higher phytate content compared to wheat gluten, soy isolate, and concentrate. The pea protein preparations had better solubility in water than other products (Table 3). The emulsifying activity in pea protein preparation was comparable to that of the soy isolate. It appears that there is no correlation between phytate and the functional performance of proteins; this again underscores the importance of type of protein(s) involved in interactions with phytate. Dev and Mukherjee[36] evaluated the functional properties of three rapeseed protein isolates containing different levels of phytate along with that of a soy protein isolate. Data on several functional characteristics are summarized in Table 4. Low-phytate products had better emulsifying properties than their high-phytate counterparts. Phytate did not significantly affect the foaming properties of rapeseed proteins.

Such phytic acid binding to proteins can occur under certain conditions; there has been a concern regarding how this binding may affect the protein digestibility as well as mineral bioavailability. de Rham and Jost[11] stated: "At the level present in soy protein products, the influence of phytate on soy protein solubility and the nutritional value of the protein itself is not measurable, and can be neglected." They found that there was no significant difference between soy protein isolate (protein efficiency ratio) (PER) values in either the presence or absence of phytate, indicating phytate did not influence the soy protein quality *in vivo*. Satterlee and Abdul-Kadir[37] also found that reducing the phytate content of soy protein isolate did not result in significant improvement in protein digestibility as determined by both *in vitro* digestibility and the rat bioassay. It was reported that if proteins are incubated with trypsin in the presence of

TABLE 4
Functional Properties of Rapeseed and Soybean Protein Products[36]

Sample product	Phytate (%)	Nitrogen solubility (pH 7.0)	Water absorption[a] (g/g)	Moisture adsorption[b] (g/g)	Oil absorption[c] (g)	Bulk density[d] (g/ml)	Least gelation concentration (%)	Emulsifying activity (pH 7.0) (%)	Emulsion stability[e] (pH 7.0) (%)	Emulsion capacity[f] (pH 7.0) ml/oil/ 100 mg sample	Foaming capacity (volume increase)[g] (%)
Rapeseed meal	4.3	40.7	2.18	8.49	1.39	0.234	10	61	89	21.7	209
Rapeseed protein isolate I	4.6	5.0	2.34	7.57	1.96	0.296	8	68	80	25.0	233
Rapeseed protein isolate II	1.3	17.6	2.10	7.23	1.92	0.273	12	61	61	30.3	214
Rapeseed protein isolate III	0.9	50.2	1.96	8.13	0.95	0.671	>16	87	91	26.2	208
Soybean meal	1.4	76.8	1.75	10.72	1.51	0.381	12	65	67	26.5	173
Soybean protein isolate	0.7	56.0	1.49	12.41	1.56	0.434	16	66	65	43.5	204

[a] Average of five determinations by centrifugal (0.5 g in 3 ml dispersions).
[b] When sample was exposed to a relative humidity of 84% in a desiccator at 25°C, average of duplicates.
[c] Corn oil was used. Average of five determinations.
[d] Average of four determinations.
[e] Used corn oil and 2.5% (w/v) dispersions. Average of duplicates.
[f] Done by titrating 10 ml of 1% (w/v) aqueous dispersion with corn oil (colored with Oil-Red-O) to the emulsion break point.
[g] Used 10% (w/v) aqueous dispersions. Whipping was for 5 min at high speed in a Virh's 23 blender. Average of duplicates.

sufficient amounts of phytate, trypsin should be inhibited and consequently there should be little or no hydrolysis of proteins.[38] To test this possibility, Reddy et al.[39] prepared a combined density fraction (CDF) from Great Northern beans by differential centrifugation. The CDF had 55.2 mg phytate/g, representing more than 80% of the total phytate. When this CDF was dialyzed at pH 4.0, 6.3, 7.0, and 9.0, cooked in water at 97°C for 30 min and then subjected to trypsin hydrolysis, it was observed that the major proteins in the CDF were completely hydrolyzed in 30 min at 37°C, regardless of the phytate:trypsin ratio in the mixture (Figure 6).

Carnovale et al.[40] studied the effect of phytic acid on the *in vitro* digestibility of several proteins (faba bean, pea, lactalbumin, casein, serum albumin, zein, and soy protein isolate) using a multienzyme method. Their data indicate that: (1) binding between phytic acid and protein is not affected by physical methods of separation (dehulling and air classification) and the weight ratio of phytic acid to protein remained constant (1:29) in different fractions of the same source in case of both faba bean and peas; (2) exogenous phytic acid (addition of sodium phytate) decreases the *in vitro* protein digestibility by 6.8, 5.7, and 8.7% in whole bean flour, protein concentrate, and protein isolate respectively; and (3) adding exogenous phytate to lactalbumin, serum albumin, casein, zein, and soy protein isolate also causes a reduction in *in vitro* digestibility of these proteins, with lowest reduction being in lactalbumin (3.8%) and the highest for zein (11.1%). Based on these findings, they concluded that the phytate-protein interactions affect protein availability negatively and that the extent of this effect depends on the nature and source of the protein.

It is apparent that there is conflicting data on the effect of phytate on protein digestibility. Factors such as polyphenolic compounds, absence/presence of minerals in the system, processing of the products, and presence/absence of fiber in such preparations undoubtedly complicates the understanding of the effect of phytate on protein digestibility.

III. PHYTATE-MINERAL INTERACTIONS

The concern over phytate-mineral interactions arises due to the ability of phytate to form insoluble complexes with minerals at physiological pH values. The main nutritional concerns associated with phytates on bioavailability of minerals (calcium, magnesium, iron, zinc, and others) from cereals, legumes, and their derived foods to humans and animals are presented in a separate chapter. The physiochemical aspects of phytate-metal interactions are discussed in this chapter.

Phytic acid, being a strong acid, forms a variety of salts with metals. Maddaiah et al.[41] and Vohra et al.[1] studied the solubility and relative stabilities of various phytate-metal complexes using potentiometric titration methods. Vohra et al.[1] indicated the order of stability as $Cu^{2+} > Zn^{2+} > Ni^{2+} > Co^{2+} > Mn^{2+} > Fe^{3+} > Ca^{2+}$ at pH 7.40; whereas Maddaiah et al.[41] found the order of stability to be $Zn^{2+} > Cu^{2+} > Co^{2+} > Mn^{2+} > Ca^{2+}$. Copper and zinc appear to have a high affinity for phytate to form complexes. They also observed that the metal binding to phytic acid is pH dependent. Solubility of these phytate-metal complexes depends on the pH and experimental conditions.

Grynspan and Cheryan[42] investigated the effects of pH and molar ratio of calcium and phytic acid on the interactions between calcium and phytic acid using atomic absorption. They concluded that: (1) calcium and phytic acid are soluble below pH 4.0 at all molar ratios, (2) above pH 4.0 the extent of solubility drop depends upon the calcium:phytic acid ratio, (3) above pH 6.0 the highest calcium precipitation occurs at calcium to phytic acid ratios of 4.0 to 6.5, (4) phytic acid solubility decreases with increasing calcium:phytic acid ratio and complete precipitation occurs at ratio of 5, and (5) when calcium was limiting, phytic acid remains in solution as the pH is increased to 7.0. Evans and Pierce[43] also found that when molar ratio of calcium to phytic acid is 5, precipitation occurs in the pH range of 5.0 to 6.0 under *in vitro* conditions at 25°C. The stability of calcium-phytate (Ca_5-phytate) complex was calculated to be as 10^{-22} in the presence of $0.2M$ KCl in the pH range of 5.0 to 6.0. The calcium-phytic acid

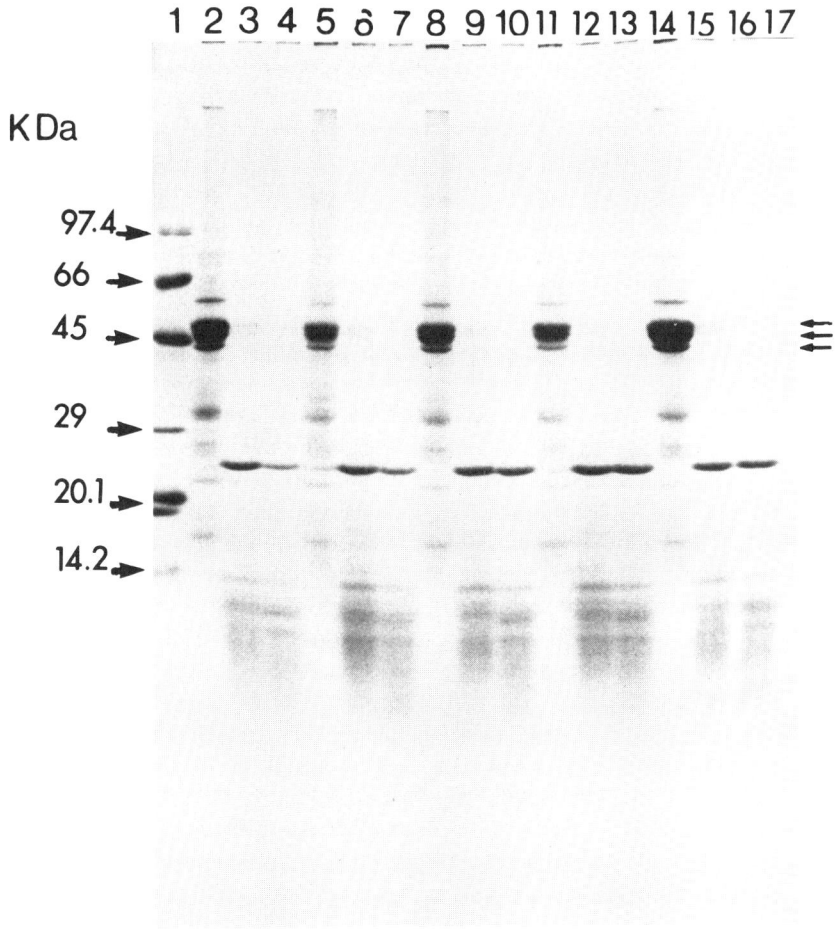

FIGURE 6. Sodium dodecyl sulfate polyacrylamide gel electrophoresis (SDS-PAGE) of trypsin-digested CDF and various dialyzed CDFs. Lane 1: molecular weight markers, phosphorylase b (97.4 kDa), carbonic anhydrase (29 kDa), soybean trypsin inhibitor (20.1 kDa), and alpha-lactalbumin (14.2 kDa). Lanes 2, 5, 8, 11, and 14, respectively: CDF, CDF (pH 4.0), CDF (pH 6.3), CDF (pH 7.0), and CDF (0.0) and no trypsin treatment. Lanes 3, 6, 9, 12, and 15, respectively: CDFs in lanes 2, 5, 8, 11 and 14 digested with trypsin for 10 min. Lanes 4, 7, 10, 13, and 17, respectively: CDFs in lanes 2, 5, 8, 11, and 14 digested with trypsin for 30 min. Lane 17: trypsin control. The weight to weight ratio between protein, phytate, and trypsin was 10:1:1 in CDF, 10:0.5:1 in dialyzed CDFs (pH 4.0, 6.3, and 9.0), and 10:0.4:1 in dialyzed CDF (pH 7.0). (From Reddy, N. R., Sathe, S. K., and Pierson, M. D., *J. Food Sci.*, 53, 107, 1988. With permission.)

interactions are reported to be endothermic and associated with a positive entropy change.[44,45] Graf[45] studied binding of calcium to phytic acid by calcium selective potentiometric methods using several pH, ionic strength, and temperature conditions. He observed existence of intrinsically different binding sites on phytic acid molecule for calcium, increase in affinity of these sites with pH, and increase in affinity of phytic acid towards calcium as the temperature increased. Further, he observed that there were three different classes of binding sites on phytic acid for calcium, resulting in two different soluble calcium-phytic acid complexes never exceeding a bound calcium concentration of 2.2 to 2.4 (moles of bound calcium per mole of phytic acid) when the binding studies were carried out at pH 4.8 and 5, 20, and 40°C. In contrast, Martin and Evans[53] found no binding of calcium to phytic acid below pH 5.0. They reported that 4.8 moles of Ca(II) was bound to phytic acid above pH 8.0 and maximum binding occurred at a Ca(II):phytic acid molar ratio of 6. Concentration of reactants and pH influence the binding between calcium and phytic acid.[45,53]

In a study on magnesium-phytic acid complexes, Cheryan et al.[48] made several observations including: (1) magnesium-phytic acid complexes are soluble at a pH below 5.0, (2) solubility of the complex decreases when the pH is more than 5.0, and solubility decreases when the magnesium:phytic acid ratio increases, (3) phosphorus is more soluble than magnesium when the magnesium:phytic acid ratio is 4 or less and pH was more than 5.0, (4) magnesium is more soluble than phosphorus when the magnesium:phytic acid ratio is more than 6.0, and (5) the penta-magnesium form of phytic acid is predominant when magnesium is in excess. This study suggests that the metal (divalent):phytic acid molar ratios cause the precipitation of the metals.

Jacobsen and Slotfeldt-Ellingsen[49] investigated the binding of copper to phytic acid using an electronic paramagnetic resonance technique at stomach pH under *in vitro* conditions. They found that the amount of copper bound to phytic acid is dependent on the type of amino acid present in the mixture. Addition of amino acids decreases the ratio of copper to phytic acid.

Accessibility of trace metals (cadmium, copper, lead, and zinc) co-precipitated with calcium phytate to soluble chelating agents was studied.[50] The accessibility ranges from 58% for zinc at 37°C to 92% for cadmium at 20°C, depending on the type of the metal, ratio of trace metal to phytate, concentration of calcium phytate, and temperature. Zinc is the most sensitive to the conditions of precipitation.

Several other studies[46,47,51-59] elucidated the thermodynamic, stoichiometric, metal competitions, and interaction aspects of metal binding to phytic acid under *in vitro* conditions. These studies provide insights into phytate-metal complexes highlighting the differences and similarities between different metals, as well as the types of products and characteristics. Some of these studies emphasize the importance of environmental manipulations with respect to pH, concentrations of interacting species, temperature, and presence and absence of other components (chelating agents or other metals) in phytate-metal complexes. In addition to these factors, the molecular configuration of phytic acid may also be one of the variables that needs to be considered in the study of phytate-metal complexes[53,56] because phytic acid exists both in axial form (pH 5.0 to 12.0) and equatorial (pH below 5.0 or above 12.0) forms. Pierce[60] indicated that to elucidate the nature of products with respect to structure is difficult because the stoichiometry and the number of products one obtains are governed by experimental conditions. Additional factors such as presence/absence of cationic/anionic species, processing conditions, and competition between these for phytic acid or metal ion would have significant effects on phytate-metal interactions in a complex food system.

REFERENCES

1. **Vohra, P., Gray, G. A., and Kratzer, F. H.,** Phytic acid-metal complexes, *Proc. Soc. Exp. Biol. Med.,* 120, 447, 1965.
2. **Barré, R., Curtois, J. E., and Wromser, G.,** Étude de la structure de l'acide phytique au moyen de ses courbes, de titration et de la conductivité de ses solutions, *Bull. Soc. Chem. Biol.,* 36, 455, 1954.
3. **Costello, A. J. R., Glonek, T., and Myers, T. C.,** Phosphorus-31 nuclear magnetic resonance-pH titrations of myoinositol hexaphosphate, *Carbohydrate Res.,* 46, 159, 1976.
4. **Evans, W. J., McCourtney, E. J., and Shrager, R. I.,** Titration studies of phytic acid, *J. Am. Oil Chem. Soc.,* 59, 189, 1982.
5. **Cheryan, M.,** Phytic acid interactions in food systems, *CRC Crit. Rev. Food Sci. Nutr.,* 13, 297, 1980.
6. **Omosaiye, O. and Cheryan, M.,** Low-phytate, full fat soy protein product by ultrafiltration of aqueous extracts of whole soybeans, *Cereal Chem.,* 56, 58, 1979.
7. **Thompson, L. U.,** Reduction of phytic acid concentration in protein isolates by aceylation techniques, *J. Am. Oil Chem. Soc.,* 64, 1712, 1987.
8. **Fontaine, T. D., Pons, W. A., and Irving, G. W.,** Protein-phytic acid relationship in peanuts and cottonseed, *J. Biol. Chem.,* 164, 487, 1946.
9. **Smith, A. K. and Rackis, J. J.,** Phytin elimination in soybean protein isolation, *J. Am. Chem. Soc.,* 79, 633, 1957.
10. **Saio, K., Koyama, E., and Watanabe, T.,** Protein-calcium-phytic acid relationships in soybean. I. Effects of calcium and phosphorus on solubility characteristics of soybean meal protein, *Agric. Biol. Chem.,* 31, 1195, 1967.
11. **de Rham, O. and Jost, T.,** Phytate-protein interactions in soybean extracts and low-phytate soy protein products, *J. Food Sci.,* 44, 596, 1979.
12. **Chen, B. H. Y. and Morr, C. V.,** Solubility and foaming properties of phytate-reduced soy protein isolate, *J. Food Sci.,* 50, 1139, 1985.
13. **Wolf, W. J. and Briggs, D. R.,** Purification and characterization of 11S component of soybean proteins, *Arch. Biochem. Biophys.,* 85, 186, 1959.
14. **Saio, K., Koyama, E., and Watanabe, T.,** Protein-calcium-phytic acid relationships in soybean. II. Effects of phytic acid on combination of calcium with soybean meal protein, *Agric. Biol. Chem.,* 32, 448, 1968.
15. **Saio, K., Koyama, E., and Watanabe, T.,** Protein-calcium-phytic acid relationships in soybeans. III. Effect of phytic acid on coagulative reaction in tofu-making, *Agric. Biol. Chem.,* 33, 36, 1969.
16. **Okubo, K., Myers, D. V., and Iacobucci, G. A.,** Binding of phytic acid to glycinin, *Cereal Chem.,* 53, 513, 1976.
17. **O'Dell, B. L. and deBoland, A.,** Complexation of phytate with proteins and cations in corn germ and oilseed meals, *J. Agric. Food Chem.,* 24, 804, 1976.
18. **Brooks, J. R. and Morr, C. V.,** Phosphorus and phytate content of soybean protein components, *J. Agric. Food Chem.,* 32, 672, 1984.
19. **Appurao, A. G. and Narasingarao, M. S.,** Binding of Ca(II) by the 11S fraction of soybean proteins, *Cereal Chem.,* 52, 21, 1975.
20. **Wallace, G. W. and Satterlee, L. D.,** Calcium binding and its effect on the properties of several food protein sources, *J. Food Sci.,* 42, 473, 1977.
21. **Prattley, C. A. and Stanley, D. W.,** Protein-phytate interactions in soybeans. I. Localization of phytate in protein bodies and globoids, *J. Food Biochem.,* 6, 243, 1982.
22. **Honig, D. H. and Wolf, W. J.,** Mineral and phytate content and solubility of soybean protein isolates, *J. Agric. Food Chem.,* 35, 583, 1987.
23. **Khan, A.,** Zinc in Soybeans: Chemical Nature and Bioavailability, Ph.D. dissertation, Purdue University, W. Lafayette, 1987.
24. **Hellot, R. and Macheboeuf, M.,** The peptides of the peanut *(Arachis hypogea),* identification of phosphate impurities, *Bull. Soc. Chim. Biol.,* 29, 817, 1947.
25. **Hill, R. and Tyler, C.,** The reaction between phytate and protein, *J. Agric. Sci.,* 44, 324, 1954.
26. **Bourdet, A. and Feillet, P.,** Distribution of phosphorus compounds in the protein fractions of various types of wheat flours, *Cereal Chem.,* 44, 457, 1967.
27. **Champagne, E. T., Rao, R. M., Liuzzo, J. A., Robinson, J. W., Gale, R. J., and Miller, F.,** Solubility behaviors of the minerals, proteins, and phytic acid in rice bran with time, temperature, and pH, *Cereal Chem.,* 62, 218, 1985.
28. **Champagne, E. T., Rao, R. M., Liuzzo, J. A., Robinson, J. W., Gale, R. J., and Miller, F.,** The interactions of minerals, proteins, and phytic acid in rice bran, *Cereal Chem.,* 62, 231, 1985.
29. **Bourdillon, J.,** A crystalline bean seed protein in combination with phytic acid, *J. Biol. Chem.,* 189, 65, 1951.
30. **Reddy, N. R. and Pierson, M. D.,** Isolation and partial characterization of phytic acid-rich particles from Great Northern beans *(Phaseolus vulgaris* L.), *J. Food Sci.,* 52, 109, 1987.

31. Reddy, N. R. and Salunkhe, D. K., Interactions between phytate, protein, and minerals in whey fractions of black gram, *J. Food Sci.*, 46, 564, 1981.
32. Barré, R. and Van Huot, N., Étude de la combinaison de l'acide phytique avec la serum-albumine humaine nativé, acetylée et déaminée, *Bull. Soc. Chim. Biol.*, 47, 1399, 1965.
33. Prattley, C. A., Stanley, D. W., and Van De Voort, F. R., Protein-phytate interactions in soybeans. II. Mechanism of protein-phytate binding as affected by calcium, *J. Food Biochem.*, 6, 255, 1982.
34. Hartman, G. H., Removal of phytate from soy protein, *J. Am. Oil Chem. Soc.*, 56, 731, 1977.
35. Naczk, M., Rubin, L. J., and Shahidi, F., Functional properties and phytate content of pea protein preparations, *J. Food Sci.*, 51, 1245, 1986.
36. Dev, D. K. and Mukherjee, K. D., Functional properties of rapeseed protein products with varying phytic acid contents, *J. Agric. Food Chem.*, 34, 775, 1986.
37. Satterlee, L. D. and Abdul-Kadir, R., Effect of phytate content on protein nutritional quality of soy and wheat bran proteins, *Lebensm.-Wiss. Technol.*, 16, 8, 1983.
38. Singh, M. and Krikarian, A. D., Inhibition of trypsin activity *in vitro* by phytate, *J. Agric. Food Chem.*, 30, 799, 1982.
39. Reddy, N. R., Sathe, S. K., and Pierson, M. D., Removal of phytate from Great Northern beans *(Phaseolus vulgaris* L.) and its combined density fraction, *J. Food Sci.*, 53, 107, 1988.
40. Carnovale, E., Lugaro, E., and Lombardi-Boccia, G., Phytic acid in faba bean and pea: effect on protein availability, *Cereal Chem.*, 64, 114, 1988.
41. Maddaiah, V. T., Kurnick, A. A., and Reid, B. L., Phytic acid studies, *Proc. Soc. Exp. Biol. Med.*, 115, 391, 1964.
42. Grynspan, F. and Cheryan, M., Calcium phytate: effect of pH and molar ratio on *in vitro* solubility, *J. Am. Oil Chem. Soc.*, 60, 1761, 1983.
43. Evans, W. J. and Pierce, A. G., Calcium-phytate complex formation studies, *J. Am. Oil Chem. Soc.*, 58, 850, 1981.
44. Evans, W. J, Marini, M. A., and Martin, C. J., Heats of precipitation of calcium phytate, *J. Inorg. Biochem.*, 19, 129, 1983.
45. Graf, E., Calcium binding to phytic acid, *J. Agric. Food Chem.*, 31, 851, 1983.
46. Martin, C. J. and Evans, W. J., Phytic acid-zinc ion interactions: a colorimetric and titrimetric study, *J. Inorg. Biochem.*, 26, 169, 1986.
47. Martin, C. J. and Evans, W. J., Phytic acid: divalent cation interactions. III. A colorimetric and titrimetric study of the pH dependence of copper (II) binding, *J. Inorg. Biochem.*, 28, 39, 1986.
48. Cheryan, M., Anderson, F. W., and Grynspan, F., Magnesium-phytate complexes: effects of pH and molar ratio on solubility characteristics, *Cereal Chem.*, 60, 235, 1983.
49. Jacobsen, J. and Slotfeldt-Ellingsen, D., Phytic acid and metal availability: a study of Ca and Cu binding, *Cereal Chem.*, 60, 392, 1983.
50. Wise, A. and Gilburt, D. J., Accessibility of trace metals, co-precipitated with calcium phytate to soluble chelating agents, *Nutr. Res.*, 3, 321, 1983.
51. Evans, W. J. and Pierce, A. G., Jr., Interaction of phytic acid with metal ions copper (II), cobalt (II), iron (III), magnesium (II), and manganese (II), *J. Food Sci.*, 47, 1014, 1982.
52. Nolan, K. B., Duffin, P. A., and McWeeny, D. J., Effects of phytate on mineral bioavailability: *in vitro* studies on Mg^{2+}, Ca^{2+}, Fe^{3+}, Cu^{2+}, Zn^{2+}, Cd^{2+} solubilities in the presence of phytate, *J. Sci. Food Agric.*, 40, 79, 1987.
53. Martin, C. J. and Evans, W. J., Phytic acid-metal ion interactions. II. The effect of pH on Ca (II) binding, *J. Inorg. Biochem.*, 27, 17, 1986.
54. Evans, W. J., Jacks, T. J., and McCourtney, E. J., The interaction of zinc ion with phytic acid, *J. Food Sci.*, 48, 1208, 1983.
55. Champagne, E. T. and Hinojosa, O., Independent and mutual interactions of copper (II) and zinc (II) ions with phytic acid, *J. Inorg. Biochem.*, 30, 15, 1987.
56. Wise, A., Influence of calcium on trace metal-phytate interactions, in *Phytic Acid: Chemistry and Applications*, Graf, E., Ed., Pilatus Press, Minneapolis, 1986, 151.
57. Wise, A. and Gilburt, D. J., *In vitro* competition between calcium phytate and the soluble fraction of rat small intestine contents for cadmium, copper, and zinc, *Toxicol. Lett.*, 11, 49, 1982.
58. Egbewatt, A. N. and Dill, K., Interaction of Mn^{2+} and Gd^{3+} with phytic acid, *Inorg. Chim. Acta*, 136, L37, 1987.
59. Martin, C. J. and Evans, W. J., Phytic acid: divalent cation interactions. V. Titrimetric, colorimetric, and binding studies with cobalt (II), nickel (II) and their comparison with other metal ions, *J. Inorg. Biochem.*, 30, 101, 1987.
60. Pierce, A. G., Jr., Structure studies of phytate-zinc ion complexes: X-ray diffraction and thermal analysis, *Inorg. Chim. Acta*, 106, L9, 1985.

Chapter 8

PHYTATE DIGESTION AND BIOAVAILABILITY

I. INTRODUCTION

Phytic acid combines with cations in the digestive tract of animals to form insoluble complexes. Calcium phytate is one of the most insoluble salts of phytate.[1] Once these complexes are formed, the phosphorus constituent cannot be utilized by animals unless phytate is hydrolyzed by phytase. The phosphorus in phytate is regarded as being less biologically available than most forms of inorganic phosphorus. The availability of phosphorus from plant sources, when present in phytate, depends on the animal species, the age of the animal, and the ability of the animal to hydrolyze phytic acid, i.e., level of phytase activity in the intestinal tract of the animal.

II. ANIMALS

Phytate is reported to be unavailable for utilization by monogastric animals but it can be hydrolyzed by ruminants due to the presence of microbial phytase in the rumen. In ruminants the bioavailability of phosphorus from plant sources is greater than 50%.[2] Reid et al.[3] reported sheep could utilize phytate and most of the phytate hydrolysis occurs in the rumen in less than 8 hours. Using an artificial rumen technique, Raun et al.[4] showed that rumen microorganisms from a steer hydrolyzed calcium phytate, suggesting the presence of phytase. Since then, several studies[2,5-9] have shown that ruminants (sheep and cattle) can utilize variable amounts of dietary phytate. Biological values for phosphorus ingested as dietary phytate in sheep and dairy cattle were reported to be 66% and 50%, respectively.[2,5] Recently, Nelson et al.[10] studied the intestinal hydrolysis of natural phytate in calves and steers that were fed a diet composed primarily of either corn or sorghum grain and soybean meal. For 9-month-old steers (average weight 200 kg) fed 71 g of phytate phosphorus, none of the phosphorus was excreted, while 56-day-old calves fed 20 g of phytate phosphorus excreted 0.06 g of the phosphorus. No phytate was recovered from the contents of the rumen, abomasum, small and large intestines of calves fed a diet composed primarily of soybean meal and sorghum grains. Further, they inferred that the initial phytate hydrolysis occurred in the rumen and was complete before the feed reached the other parts of the digestive system.

About 98% of dietary phytate phosphorus is hydrolyzed to inorganic phosphorus during early lactation of Holstein cows fed a diet containing large amounts of grain.[11] Neither the source nor the amount of calcium is reported to have a significant effect on phytate hydrolysis in dairy cows.

Nelson et al.[1] investigated the hydrolysis of phytate by rabbits (pseudoruminants), which were fed one of four dietary treatments. The dietary treatments were blends consisting of 60% basal diet plus 40% corn or wheat shorts or 75% basal diet plus 25% soybean meal or wheat bran. The phytate content in these four dietary treatments ranged from 1.77 to 4.19 mg/g. For all treatments no phytate was excreted. This complete hydrolysis of phytate was probably due to bacterial phytase in the cecum.

Swine, as monogastrics, can utilize variable amounts of phytate phosphorus. Based on a summary of several studies (Table 1), the availability of phosphorus from phytate phosphorus ranges from 20 to 60%, with an average value of 33% for pigs 50 to 90 lb in weight. There is some indication that the ability of swine to utilize phytate phosphorus improves with age. Tonroy et al.[16] reported that the apparent digestibility of phosphorus in sorghum grain and soybean meal is inferior to that of dicalcium phosphate when fed to growing swine. A recent study of Calvert et al.[17] also showed that the natural phytate from barley and corn is poorly digested by growing swine. Pierce et al.[18] studied the availability of phytate phosphorus to

TABLE 1
Biological Value of Phytate Phosphorus in Swine

Pig weight (lbs)	Biological availability (%)	Ref.
60	20 to 30	12
50—90	30 to 40	15
50	18 to 24	13
Growing	30 to 60	14

growing pigs (11 to 14 kg in weight) fed wheat and/or corn-based diets. They concluded that growing pigs were able to develop normally when fed wheat or corn-based diets containing 0.65% calcium and 0.50% total phosphorus, with up to 0.30% as phytate phosphorus. Overall performance and development was impaired when pigs were fed similar diets containing 0.38% phytate. Recent studies from the People's Republic of China have shown that pigs utilize up to 37% of the phytate phosphorus from the common Chinese feedstuffs.[19]

The type of cereal in a diet also influences phytate phosphorus utilization in swine. Pointillart et al.[20] found that the phosphorus utilization by pigs fed wheat was 1.7 times higher than that of corn-fed pigs. Both wheat-based and corn-based diets contained equal amounts of total phosphorus and phytate phosphorus. The better utilization of phosphorus by pigs fed a wheat-based diet may be due to a higher phytase activity in wheat grain, since the intestinal phytase activity of pigs fed on wheat- and corn-based diets was the same.

One study has suggested that dietary vitamin D enhances phytate phosphorus utilization.[21] Vitamin D improves phytate phosphorus absorption through a mechanism which does not involve an increase in the activity of the intestinal phosphatases. However, two intestinal vitamin D sensitive enzymes, alkaline phosphatase and phytase, are thought to be involved in phytate phosphorus digestion.

III. POULTRY

Phytate in cereal grains and other plant foodstuffs has been reported to be poorly available to chicks when supplied either in its natural form or extracted as the calcium or sodium salts.[22,23,80] Extensive studies have been done on the availability of phosphorus from phytate phosphorus to poultry. Heuser et al.,[24] McGinnis et al.,[25] Singsen et al.,[26] Gillis et al.,[27] and Sunde and Bird[28] found that natural phytate is a poor source of phosphorus for various species of poultry. In contrast, Sieburth et al.[29] reported that the phosphorus in finely ground whole wheat flour was almost completely available to chicks for growth but was less available than inorganic phosphate for bone deposition. Temperton et al.[30-32] concluded that pullet chicks (less than four weeks old), growing pullets (reared to 18 weeks of age), and laying hens were able to utilize effectively the organic sources of phosphorus for growth and bone formation.

Several investigators have fed poultry with various isolated phytate compounds as a source of phosphorus. Lowe et al.[33] reported that chicks did not efficiently utilize isolated phytate phosphorus from wheat bran. Singsen and Mitchell[34] and Matterson et al.[35] found calcium-magnesium phytate to be a poor source of phosphorus for the turkey poult. Gillis et al.[36] showed that chicks were unable to utilize relatively pure calcium phytate. Conversely, Harms et al.[37] and Waldroup et al.[38] concluded that the phosphorus in phytic acid was highly available to the chick. Gillis et al.[39] studied the quantitative utilization of phytate phosphorus by white leghorn hens. They found phosphorus from isolated calcium phytate was biologically less available (less than 50%) than from dicalcium phosphate as indicated by mortality, egg production, and bone

TABLE 2
Biological Value of Phytate Phosphorus (Calcium Phytate) for Laying Hens

Biological availability (%)	Ref.
50	39
30	41
80	40
54	Average (of studies)

mineral changes. Others[40,41] concluded that phosphorus from calcium phytate is available (between 30 and 80%) to laying hens. A summary of availability of phosphorus from calcium phytate for the laying hen from several studies is presented in Table 2.

Few investigators have studied the quantitative biological utilization of phytate phosphorus by chicks and turkeys. Gillis et al.[42] fed chicks and turkeys ^{32}P-labeled calcium phytate and ^{32}P-labeled monosodium orthophosphate and then measured the amount of radioactivity retained in the tibia. The chicks used only 10% of phosphorus from calcium phytate and the corresponding utilization of calcium phytate by the turkey was less than 2%. In another study, Ashton et al.[43] fed ^{32}P-labeled calcium phytate and observed that four-week-old chicks retained approximately 20% of phytate compared to six-week-old chicks that retained 36 to 49% of phytate consumed. Temperton and Cassidy[44,45] reported that chicks utilized phytate from foodstuffs of plant origin for deposition in the growing bones and retained approximately 60% of the phytate phosphorus in their body. Andrews et al.[46,47] concluded that the phosphorus in all-plant phosphorus diet containing hominy feed appeared to be 60 to 80% available to the turkey poult. Recent studies by Su et al.[19] and Yu et al.[48] show that the chicks utilize only a small amount (7%) of phytate phosphorus from Chinese feedstuffs that contained 20 to 80% of the phosphorus as phytate. They observed a strong negative correlation between phytate phosphorus in the diet and absorption of phosphorus.

Nelson et al.[49] studied the effect of supplemental phytase on the utilization of phosphorus from phytate by chicks. Phytase was prepared as an acetone-dried powder from the culture fluid of *Aspergillus ficuum* NRRL 3135 and was added to the diet at levels up to 3g/1000 g. At this level, chicks utilized phosphorus from phytate of diet as efficiently as supplemental inorganic phosphate. The phytase added to the diet was active in the alimentary tract of the chick and not in the feed prior to ingestion. A study by Mohammed et al.[50] indicated that the digestibility of phytate was higher when dietary (plant) phytase was present in the diet regardless of the bacteriological status of the chick gut. They concluded that the intestinal microflora had little or no role in the hydrolysis of dietary phytate in the chick.

The ability of young chicks to hydrolyze phytate appears to increase with age up to maturity.[51,81] Nelson[51] measured the amount of natural phytate hydrolyzed by chicks and laying hens. Chicks (four weeks old), chicks (nine weeks old), and laying hens, hydrolyzed, respectively, 0, 3, and 8% of the natural phytate, when the diet contained corn as the only grain source (Table 3). On the other hand, four-week-old chicks, nine-week-old chicks, and laying hens hydrolyzed, respectively, 8, 13, and 13%, of the natural phytate, when 50% of corn was replaced by wheat in the diet. It appears that the laying hens showed little improvement over young chicks in their ability to hydrolyze natural phytate from corn-based diets. Adverse effects of phytic acid from soybean, palm kernel, cotton seed, and rape seed meals on phosphorus retention in growing chicks has also been reported.[52]

The level of calcium and non-phytate P in the diet influences the availability of phytate

TABLE 3
Average Phytate Hydrolysis by Chicks and Laying Hens[51]

	Phytate hydrolyzed (%)	
Poultry	Diet 1	Diet 2
Chicks (4 weeks old)	0 ± 0.9	8 ± 0.8
Chicks (9 weeks old)	3 ± 1.0	13 ± 0.5
Laying Hens	8 ± 1.7	13 ± 1.2

Note: Diet 1 consisted of 54.05% ground yellow corn, 37.00% soybean meal (49% protein), 5.0% soybean oil, and other salt and vitamin mixtures. Diet 2 consisted of 27.025% wheat, 27.025% ground yellow corn, 37.0% soybean meal, 5% soybean oil, and other salt and vitamin mixtures.

TABLE 4
Effect of Dietary Levels of Calcium and Non-Phytate Phosphorus on Phytate Hydrolysis by Chicks[55]

Nonphytate phosphorus (%)	Calcium (%)	Phytate hydrolysis (%)
0.12	0.09	42.1
0.12	1.00	8.3
0.45	0.09	56.8
0.45	1.00	5.9

Note: Basal diet is composed of yellow corn, soybean meal, vitamin mixture, and trace mineral mixture.

phosphorus to chicks.[53,54,81,82] Calcium reduces the availability of phytate phosphorus to chicks by interacting with phytate to form an insoluble calcium-phytate complex.[81,82] Bellam et al.[55] studied the effect of different dietary levels of calcium and non-phytate phosphorus on the ability of chicks to hydrolyze phytate. Broiler chicks (three weeks old) were fed diets composed primarily of corn and soybean meal containing varying amounts of calcium (0.09 or 1.00%) and non-phytate phosphorus (0.12 or 0.45%). They found that increasing the calcium content of the diet to 1.00% reduced phytate hydrolysis regardless of non-phytate phosphorus content (Table 4); in the absence of added calcium, increasing the non-phytate phosphorus content to 0.45% improved phytate hydrolysis. The increase in phytate hydrolysis achieved by the addition of non-phytate phosphorus, in the absence of supplemental calcium, may have resulted from interaction of cations, which would otherwise form insoluble salts with phytate.

As discussed earlier in this section, a wide disagreement exists between investigators on the ability of poultry to utilize phytate. The disagreements could be due to variations in experimental methods and materials. These variables include species differences, age of the test animals, the source of phytate, criteria of response, and the level of calcium and vitamin D used in the experimental diets.

IV. RATS

Rats effectively utilize phytate phosphorus. However, factors such as levels of phytate and calcium in the diet may influence phytate hydrolysis in rats. Ranhotra et al.[56] fed rats bread-based

TABLE 5
Availability of Phytate to Rats[56]

Diet[a]	Phytate phosphorus hydrolyzed	
	(mg)	(%)
A (44.8)	78.6	21.8
B (179.2)	1,220.0	75.3
C (358.4)	2,391.7	83.8
D (537.6)	3,351.1	81.1

[a] Values in the parentheses refer to the amount (mg/100 g) of dietary phytate phosphorus.

TABLE 6
Effect of Diet Composition on the Hydrolysis of Phytate by Weanling and Mature Rats[63]

Diet	Phytate hydrolyzed (%)	
	Weanling rats	Mature rats
Basal diet[a]	71	27
Wheat[b]	78	41

[a] Basal diet contained 71.72% of ground yellow corn, and 23.0% soybean meal (49% protein).
[b] Wheat based diet contained 35.86% of ground yellow corn, 35.86% wheat, and 23.0% soybean meal.

diets containing increasing levels of phytate for six weeks and then measured the availability of phytate phosphorus. They found that the amount of phytate hydrolyzed increased with increased levels of dietary phytate (Table 5).

It appears that the gut microflora are an important source of phytase activity and are responsible for phytate hydrolysis in rats.[57-61] Rats adapt to low phosphorus diets by an increase in phytate digestion.[62] The adaptation may result from increased synthesis of phytase or alkaline phosphatase by the intestinal microflora when there is a low level of phosphorus in the diet.

Age and sex both affect *in vivo* phytate hydrolysis in rats.[57,63,64] Nelson and Kirby[63] observed that weanling and mature rats hydrolyzed 71 and 39%, respectively, of the phytate in a corn-soybean meal based diet. They noted that the dietary variables in the basal diet influenced the amount of phytate hydrolyzed by both weanling and mature rats. Replacing one half of the corn in the corn-soybean meal based diet with wheat increased phytate hydrolysis in both weanling and mature rats (Table 6). In a subsequent study, Bellam et al.[54] found that rats fed a diet containing wheat bran hydrolyzed more phytate (40%) than those fed other diets (corn-soybean meal based diet or a diet containing cottonseed hulls) (14.3 to 29.0%). The increased phytate hydrolysis in rats fed a diet containing wheat bran can be attributed to the presence of high phytase activity in wheat bran. A study from People's Republic of China indicated that the rats utilize 44% of the phytate from common Chinese feedstuffs.[19]

A higher dietary concentration of calcium reduces phytate hydrolysis in the rat.[59,65,55,79] Wise[67] found that a considerable proportion of the phytate in feed containing a calcium to phytate molar ratio in excess of 6:1 was hydrolyzed in the intestine of rats by the intestinal microflora but not in germ-free rats. Taylor and Coleman[65] reported that availability of phytate to rats was increased significantly as the dietary level of phytate increased and as the dietary level of calcium decreased. Increasing the calcium to phytate molar ratio above 6:1 reduces the proportion of phytate hydrolyzed in rats.[63,65] The excess calcium reduces phytate hydrolysis in rats by decreasing microbial phytase activity.[59,66] Addition of a yeast culture or yogurt to a diet composed of plant feedstuffs improves utilization of phytate in rats.[68-70] Toleman et al.[70] reported that the addition of yogurt, active or inactive, to a nutrient-balanced diet containing a high concentration of phytate, reduced the negative effects that phytate normally has on rat growth.

V. HUMANS

Although experiments to determine the bioavailability of phosphorus from legume phytates in humans has not been performed, it is estimated that 40 to 80% of the phytate in cereals and cereal products is available to man.[71-73,83,84] However, there are some reports on phytate

TABLE 7
Effect of Level of Calcium Intake on Dietary Phytate Hydrolysis in Men[78]

Daily intakes (mg)		Phytate	
Calcium	Phytate	Excreted (mg/day)	Hydrolyzed (%)
550	1800	947 ± 248	47 ± 14
1050	1800	1325 ± 135	26 ± 7
1550	1800	1500 ± 91	16 ± 5

digestion in humans. Subrahmanyan et al.[74] reported that 85% of phytate in ragi (a cereal) was hydrolyzed during digestion in humans. The findings of Sandberg et al.[75] indicated that the hydrolysis of phytate by human phytase or alkaline phosphatase does occur in the stomach and small intestine, and is not due to microbial phytase activity.

Certain methods of food processing alter phytate in foods, resulting in reduced phytate digestibility of dietary phytate in humans.[76,83] Sandberg et al.[76] investigated the effect of extrusion cooking of a high fiber cereal product on digestibility of phytate in ileostomy patients. Phytate in the extruded product (prepared using mild extrusion conditions) was much less digestible in the stomach and small intestine than the phytate contained in raw ingredients. Two possible explanations for the reduced phytate digestibility associated with the extruded product are: (1) qualitative changes in phytate, i.e., formation of undigestible phytate complexes during extrusion cooking in the product and (2) loss of phytase activity in the extruded product.[83] Also, lower inositol phosphates (inositol tri-, tetra-, and pentaphosphates) form in the foods during extrusion cooking which may not be digested by the intestinal enzymes of humans.[76] The effect of food processing on the formation of phytate degradation products and on the digestibility of phytate in humans should be further studied.[83,84]

Diets containing excess calcium relative to phytate results in a reduction in phytate utilization by humans.[83] Walker et al.[77] reported that an average of 59% of the phytate was hydrolyzed in subjects consuming a diet with high calcium and low phytate. Subjects consuming a diet with high phytate and adequate levels of calcium (500 to 600 mg/day) hydrolyzed an average of 84% phytate. Ellis et al.[78] studied the effect of the level of calcium intake on *in vivo* hydrolysis of dietary phytate in healthy men. The menus consisted of foods routinely consumed in the U.S. The intake of all nutrients except calcium was the same for three levels of calcium intakes. Mean phytate hydrolysis was 47, 26, and 16% when calcium intakes were 550, 1050, and 1550 mg, respectively (Table 7). It appears that a relationship exists between calcium intake and phytate hydrolysis in humans.

REFERENCES

1. **Nelson, T. S., Daniels, L. B., Shriver, L. A., and Kirby, L. K.,** Hydrolysis of phytate phosphorus by young rabbits, *Arkansas Farm Res.,* 34, 8, 1985.
2. **Lofgreen, G. P.,** The availability of the phosphorus in dicalcium phosphate, bone meal, soft phosphate, and calcium phytate for mature wethers, *J. Nutr.,* 70, 58, 1960.
3. **Reid, R. L., Franklin, M. C., and Hallsworth, E. G.,** The utilization of phytate phosphorus by sheep, *Austr. Vet. J.,* 23, 136, 1947.
4. **Raun, A., Cheng, E., and Burroughs, W.,** Phytate phosphorus hydrolysis and availability to rumen microorganisms, *J. Agric. Food Chem.,* 4, 869, 1956.

5. **Mathur, M. L.,** Assimilation of phytin phosphorus by dairy cows, *Indian J. Vet. Sci.*, 23, 243, 1951.
6. **Plumlee, M. P., Kennington, M. H., and Beeson, W. M.,** Utilization of phosphorus from various sources by growing-fattening swine, *J. Anim. Sci.*, 14, 1220, 1955.
7. **Tillman, A. D. and Brethour, J. R.,** Utilization of phytin phosphorus by sheep, *J. Anim. Sci.*, 17, 104, 1958.
8. **Ellis, L. C. and Tillman, A. D.,** Utilization of phytin phosphorus in wheat bran by sheep, *J. Anim. Sci.*, 20, 606, 1961.
9. **Wilson, W. M. D.,** Calcium Phytate as a Source of Phosphorus for ruminants, Ph.D. dissertation, University of Illinois, Urbana-Champaign, 1975.
10. **Nelson, T. S., Daniels, J. B., Hall, J. R., and Shields, L. G.,** Hydrolysis of natural phytate phosphorus in the digestive tract of calves, *J. Anim. Sci.*, 42, 1509, 1976.
11. **Clark, W. D., Jr., Wohlt, J. E., Gilbreath, R. L., and Zajac, P. K.,** Phytate phosphorus intake and disappearance in the gastrointestinal tract of high producing dairy cows, *J. Dairy Sci.*, 69, 3151, 1986.
12. **Bayley, H. S. and Thompson, R. G.,** Phosphorus requirements of growing pigs and effect of steam pelleting on phosphorus availability, *J. Anim. Sci.*, 28, 484, 1969.
13. **Besecker, R. J., Jr., Plumlee, M. P., Pickett, R. A., and Conrad, J. H.,** Phosphorus from barley grain for growing swine, *J. Anim. Sci.*, 26, 1477, 1967.
14. **Noland, P. R., Funderburg, M., and Johnson, Z.,** Phosphorus availability in a practical diet for swine, *J. Anim. Sci.*, 27, 1155, 1968.
15. **Woodman, H. E. and Evans, R. E.,** Nutrition of the bacon pig. XIII. The minimum level of protein intake consistent with the maximum rate of growth, *J. Agric. Sci.*, 38, 354, 1948.
16. **Tonroy, B., Plumlee, M. P., Conrad, J. H., and Cline, T. R.,** Apparent digestibility of the phosphorus in sorghum grain and soybean meal for growing swine, *J. Anim. Sci.*, 36, 669, 1973.
17. **Calvert, C. C., Besecker, R. J., Plumlee, M. P., Cline, T. R., and Forsyth, D. M.,** Apparent digestibility of phosphorus in barley and corn for growing swine, *J. Anim. Sci.*, 47, 420, 1978.
18. **Pierce, A. B., Doige, C. E., Bell, J. M., and Owen, B. D.,** Availability of phytate phosphorus to the growing pig receiving isonitrogenous diets based on wheat or corn, *Can. J. Anim. Sci.*, 57, 573, 1977.
19. **Su, Q., Yu, S. X., Duan, Y. Q., Lu, Z. H., and Liu, C. H.,** *Chinese Agric. Sci.*, 2, 75, 1982.
20. **Pointillart, A., Fontaine, N., and Thomasset, M.,** Phytate phosphorus utilization and intestinal phosphatases in pigs fed low phosphorus: wheat or corn diets, *Nutr. Rep. Intern.*, 29, 473, 1984.
21. **Fontaine, N., Fourdin, A., and Pointillart, A.,** Effects of vitamin D on intestinal phytase and alkaline phosphatase and its relationship with phytate phosphorus absorption in pigs, *Reprod. Nutr. Develop.*, 25, 717, 1985.
22. **Peeler, H. T.,** Biological availability of nutrients in seeds: availability of major mineral ions, *J. Anim. Sci.*, 35, 695, 1972.
23. **Nelson, T. S.,** The utilization of phytate phosphorus by poultry — a review, *Poultry Sci.*, 46, 862, 1967.
24. **Heuser, G. F., Norris, L. C., McGinnis, J., and Scott, M. L.,** Further evidence of the need for supplementing soybean meal chick rations with phosphorus, *Poultry Sci.*, 22, 269, 1943.
25. **McGinnis, J., Norris, L. C., and Heuser, G. F.,** Poor utilization of phosphorus in cereals and legumes by chicks for bone development, *Poultry Sci.*, 23, 157, 1944.
26. **Singsen, E. P., Matterson, L. D., and Scott, H. M.,** Phosphorus in poultry nutrition. III. The relationship between the source of vitamin D and the utilization of cereal phosphorus by the poult, *J. Nutr.*, 33, 13, 1947.
27. **Gillis, M. B., Norris, L. C., and Heuser, G. F.,** The effect of phytin on the phosphorus requirement of the chick, *Poultry Sci.*, 28, 283, 1949.
28. **Sunde, M. L. and Bird, H. R.,** A critical need of phosphorus for the young pheasant, *Poultry Sci.*, 35, 424, 1956.
29. **Sieburth, J. F., McGinnis, J., Wahl, T., and McLaren, B. A.,** The availability of the phosphorus in unifine flour for the chick, *Poultry Sci.*, 31, 813, 1952.
30. **Temperton, H. F., Dudley, J., and Pickering, G. J.,** Phosphorus requirements of poultry. IV. The effects on growing pullets of feeding diets containing no animal protein or supplementary phosphorus, *Br. Poultry Sci.*, 6, 125, 1965.
31. **Temperton, H. F., Dudley, J., and Pickering, G. J.,** Phosphorus requirements of poultry. V. The effects during the subsequent laying year of feeding growing diets containing no animal protein or supplementary phosphorus, *Br. Poultry Sci.*, 6, 135, 1965.
32. **Temperton, H. F., Dudley, J., and Pickering, G. J.,** Phosphorus requirements of poultry. VI. The phosphorus requirements of growing pullets between 8 and 18 weeks of age, *Br. Poultry Sci.*, 6, 143, 1965.
33. **Lowe, J. T., Steenbock, H., and Krieger, C. H.,** Cereals and rickets. IX. The availability of phytin P to the chick, *Poultry Sci.*, 18, 40, 1939.
34. **Singsen, E. P. and Mitchell, H. H.,** Phosphorus in poultry nutrition. I. The relation between phytin and different sources of vitamin D, *Poultry Sci.*, 24, 479, 1945.
35. **Matterson, L. D., Scott, H. H., and Singsen, E. P.,** The influence of sources of phosphorus on relative efficiency of vitamin D_3 and cod liver oil in promoting calcification in poults, *J. Nutr.*, 31, 599, 1946.

36. **Gillis, M. B., Norris, L. C., and Heuser, G. F.,** The utilization by the chick of phosphorus from different sources, *J. Nutr.,* 35, 195, 1948.
37. **Harms, R. H., Waldroup, P. W., Shirley, R. L., and Ammerman, G. B.,** Availability of phytic acid phosphorus for chicks, *Poultry Sci.,* 41, 1189, 1962.
38. **Waldroup, P. W., Ammerman, G. B., and Harms, R. H.,** The availability of phytic acid phosphorus for chicks. II. Comparison of phytin phosphorus sources, *Poultry Sci.,* 43, 426, 1964.
39. **Gillis, M. B., Norris, L. C., and Heuser, G. F.,** Phosphorus metabolism and requirements of hens, *Poultry Sci.,* 32, 977, 1953.
40. **Singsen, E. P., Matterson, L. D., Tlustohowicz, J. J., and Pudelkiewicz, W. J.,** Phosphorus in the nutrition of the adult hen. II. The relative availability of phosphorus from several sources of caged layers, *Poultry Sci.,* 48, 387, 1969.
41. **Waldroup, P. W., Simpson, C. F., Damron, B. L., and Harms, R. H.,** The effectiveness of plant and inorganic phosphorus in supporting egg production in hens and hatchability and bone development in chick embryos, *Poultry Sci.,* 46, 659, 1967.
42. **Gillis, M. B., Keane, K. W. and Collins, R. A.,** Comparative metabolism of phytate and inorganic P^{32} by chicks and poults, *J. Nutr.,* 78, 155, 1957.
43. **Ashton, W. M., Evans, C., and Williams, P. C.,** Phosphorus compounds of oats. II. Utilization of phytate phosphorus by growing chicks, *J. Sci. Food Agric.,* 11, 722, 1960.
44. **Temperton, H. F. and Cassidy, J.,** Phosphorus requirements of poultry. I. The utilization of phytin phosphorus by the chick as indicated by balance experiments, *Br. Poultry Sci.,* 5, 75, 1964.
45. **Temperton, H. F. and Cassidy, J.,** Phosphorus requirements of poultry. II. The utilization of phytin phosphorus by the chick for growth and bone formation, *Br. Poultry Sci.,* 5, 81, 1964.
46. **Andrews, T. L., Damron, B. L., and Harms, R. H.,** Utilization of plant phosphorus by the turkey poult, *Poultry Sci.,* 51, 1248, 1972.
47. **Andrews, T. L., Damron, B. L., and Harms, R. H.,** Utilization of various sources of plant phosphorus by the turkey poult, *Nutr. Rep. Intern.,* 6, 251, 1972.
48. **Yu, S. X., Su, Q., Duan, Y. Q., Lu, Z. H., and Liu, J. X.,** Determination of utilization rate of phosphorus in phytate by growing chicks, *Chinese J. Anim. Sci.,* 4, 8, 1983.
49. **Nelson, T. S., Shieh, T. R., Wodzinski, R. J., and Ware, J. H.,** Effect of supplemental phytase on the utilization of phytate phosphorus by chicks, *J. Nutr.,* 101, 1289, 1971.
50. **Mohammed, A., Grimble, R. F., and Taylor, T. G.,** A study of the possible rate of the intestinal microflora in phytate hydrolysis in chicks, *Proc. Nutr. Soc.,* 45, 117A, 1986.
51. **Nelson, T. S.,** The hydrolysis of phytate phosphorus by chicks and laying hens, *Poultry Sci.,* 55, 2262, 1976.
52. **Nwokolo, E. N. and Bragg, D. B.,** Influence of phytic acid and crude fibre on the availability of minerals from four protein supplements in growing chicks, *Can. J. Anim. Sci.,* 57, 475, 1977.
53. **Nott, H., Morris, T. R., and Taylor, T. G.,** Utilization of phytate phosphorus by laying hens and young chicks, *Poultry Sci.,* 46, 1301, 1967.
54. **Bellam, G. C., Nelson, T. S., and Kirby, L. K.,** Effect of fiber and phytate source and of calcium and phosphorus level on phytate hydrolysis in the chick, *Poultry Sci.,* 63, 333, 1984.
55. **Bellam, G. C., Nelson, T. S., and Kirby, L. K.,** Effect of different dietary levels of calcium and phosphorus on phytate hydrolysis by chicks, *Nutr. Rep. Intern.,* 32, 909, 1985.
56. **Ranhotra, G. S., Loewe, R. J., and Puyat, L. V.,** Effect of dietary phytic acid on the availability of iron and phosphorus, *Cereal Chem.,* 51, 323, 1974.
57. **Wise, A. and Gilbert, D. J.,** Phytate hydrolysis by germ-free and conventional rats, *Appl. Environ. Microbiol.,* 43, 753, 1982.
58. **Wise, A., Richards, C. P., and Timble, M. L.,** Phytate hydrolysis in rat gastrointestinal tracts, as observed by ^{31}P Fourier Transform Nuclear Magnetic Resonance Spectroscopy, *Appl. Environ. Microbiol.,* 45, 313, 1983.
59. **Wise, A.,** Dietary calcium influences the location and extent of phytate hydrolysis in the rat intestine, in *Trace Elements in Man and Animals — TEMA-5,* Mills, C. F., Bremner, I., and Chesters, J. K., Eds., Commonwealth Agricultural Bureaux, Aberdeen, Scotland, 1985, 468.
60. **Yoshida, T. and Ohkubo, M.,** Role of gastrointestinal microflora on digestibility in young rats fed diets containing sodium phytate, *Agric. Biol. Chem.,* 48, 2571, 1984.
61. **Williams, P. J. and Taylor, T. G.,** A comparative study of phytate hydrolysis in the gastrointestinal tract of the golden hamster *(Mesocricetus auratus)* and the laboratory rat, *Br. J. Nutr.,* 54, 429, 1985.
62. **Moore, R. J. and Veum, T. L.,** Adaptive increase in phytate digestibility by phosphorus-deprived rats and the relationship of intestinal phytase and alkaline phosphatase to phytate utilization, *Br. J. Nutr.,* 49, 145, 1983.
63. **Nelson, T. S. and Kirby, L. K.,** Effect of age and diet composition on the hydrolysis of phytate phosphorus by rats, *Nutr. Rep. Intern.,* 20, 729, 1979.
64. **Yoshida, T., Shinoda, S., and Nagata, M.,** Effect of age and dietary phytate on the availability of phytate and minerals in rats, *Agric. Biol. Chem.,* 47, 2641, 1983.

65. **Taylor, T. G. and Coleman, J. W.**, A comparative study of the absorption of calcium and the availability of phytate phosphorus in the golden hamster *(Mesocricetus auratus)* and the laboratory rat, *Br. J. Nutr.,* 42, 113, 1979.
66. **Wise, A.**, Influence of calcium on trace metal-phytate interactions, in *Phytic Acid: Chemistry and Applications,* Graf, E., Ed., Pilatus Press, Minneapolis, 1986, 151.
67. **Wise, A.**, Dietary factors determining the biological activities of phytate, *Nutr. Abstr. Rev.,* 53, 791, 1983.
68. **Moore, R. J. and Veum, T. L.**, Effect of dietary phosphorus and yeast culture level on the utilization of phytate phosphorus by the rat, *Nutr. Rep. Intern.,* 25, 221, 1982.
69. **Moore, R. J. and Veum, T. L.**, Effect of source and level of dietary yeast product on phytate phosphorus utilization by rats fed low phosphorus diets, *Nutr. Rep. Intern.,* 27, 1267, 1983.
70. **Toleman, J., Stuart, M. A., and McClain, C. J.**, The effect of yogurt on growth of rats fed diets high in phytic acid, *Fed. Proc.,* 46, 1646 (abstr.), 1987.
71. **McCance, R.A. and Widdowson, E. M.**, Phytin in human nutrition. *Biochem. J.,* 29B, 2694, 1935.
72. **McCance, R. A. and Widdowson, E. M.**, Mineral metabolism of healthy adults on white and brown bread varieties, *J. Physiol.,* 101, 44, 1942a.
73. **McCance, R.A. and Widdowson, E. M.**, Mineral metabolism on dephytinized bread, *J. Physiol.,* 101, 304, 1942b.
74. **Subrahmanyan, V., Rao, M. N., Rao, G. R., and Swaminathan, M.**, The metabolism of nitrogen, calcium, and phosphorus in human adults on a poor vegetarian diet containing ragi *(Elusine coracana), Br. J. Nutr.,* 9, 350, 1955.
75. **Sandberg, A. S., Hasselblad, C., Hasselblad, K., and Hulten, L.**, The effect of wheat bran on the absorption of minerals in the small intestine, *Br. J. Nutr.,* 48, 185, 1982.
76. **Sandberg, A. S., Andersson, H., Kivisto, B., and Sandstrom, B.**, Extrusion cooking of a high-fibre cereal product. I. Effects on digestibility and absorption of protein, fat, starch, dietary fibre, and phytate in the small intestine, *Br. J. Nutr.,* 55, 245, 1986.
77. **Walker, A. R. P., Fox, F. W., and Irving, J. T.**, Studies in human mineral metabolism: The effect of bread rich in phytate phosphorus on the metabolism of certain mineral salts with special references to calcium, *Biochem. J.,* 42, 152, 1948.
78. **Ellis, R., Morris, E. R., Hill, A. D., Anderson, H. L., and McCarron, P. B.**, Effect of level of calcium intake on *in vitro* hydrolysis of dietary phytate, *Fed. Proc.,* 45, 374 (abstr.), 1986.
79. **Nahapetian, A. and Young, V. R.**, Metabolism of ^{14}C-phytate in rats: effect of low and high dietary calcium intakes, *J. Nutr.,* 110, 1458, 1980.
80. **Anon.**, Phosphorus bioavailability in poultry nutrition, *Nutr. Rev.,* 42, 387, 1984.
81. **Scheideler, S. E. and Sell, J. F.**, Utilization of phytate phosphorus in laying hens as influenced by dietary phosphorus and calcium, *Nutr. Rep. Intern.,* 35, 1073, 1987.
82. **Nelson, T. S. and Kirby, L. K.**, The calcium binding properties of natural phytate in chick diets, *Nutr. Rep. Intern.,* 35, 949, 1987.
83. **Sandberg, A. S., Andersson, H., Carlsson, N. -G., and Sandstrom, B.**, Degradation products of bran phytate formed during digestion in the human small intestine: Effect of extrusion cooking on digestibility, *J. Nutr.,* 117, 2061, 1987.
84. **Sandberg, A. S. and Andersson, H.**, Effect of dietary phytase on the digestion of phytate in the stomach and small intestine of humans, *J. Nutr.,* 118, 469, 1988.

Chapter 9

NUTRITIONAL CONSEQUENCES OF PHYTATES

I. INTRODUCTION

In studies done on animals, phytic acid in plant foods has been shown to complex with minerals such as calcium, zinc, iron, and magnesium, making these minerals unavailable for absorption.[1-4] The mechanism by which phytate affects mineral bioavailability is not clearly understood. However, some of the studies[1,5-9] have shown that the formation of insoluble phytate-mineral complexes in the intestinal tract prevents mineral absorption. The formation of these complexes is pH dependent. Data from several reports[10-16] indicate that cereal diets, legume diets, and isolated phytate products from plant sources interfere with mineral utilization. In most cases, the bioavailability of minerals has been examined in relation to the administration of free ionic salts. Results obtained under such conditions may not represent the true bioavailability of minerals from similar sources to humans.

The reduced bioavailability of minerals from legumes, cereals, and other protein foods to humans and animals due to phytate or phytate-protein complexes depends on several factors: nutritional status of animals (man); concentration of minerals and phytate in foodstuff; ability of endogenous carriers in the intestinal mucosa to absorb essential minerals bound to phytate and other dietary substances; digestion or hydrolysis of phytate by phytase and/or phosphatase enzyme in the intestine; processing of products or methods of processing which include unit food processing operations (pH adjustment; level of refinement and addition or removal of inhibitors/enhancers); and digestibility of the foodstuff.[15,17]

Other food components such as dietary fiber, polysaccharides, oxalates, and polyphenolic compounds play a major role in mineral bioavailability. For instance, dietary fiber in whole wheat bread accounts for most of the poor bioavailability of minerals.[18-21] Conversely, Davies et al.[22] suggested that phytate rather than fiber largely determines the availability of zinc for absorption. Davies and co-workers[22-25] published several reports dealing with the bioavailability of minerals from high phytate and high fiber cereals and legumes. The role of dietary fiber in mineral utilization is not fully understood.[16] Although there is considerable potential for increased utilization of cereal and legume-based foods in developed countries, and an already existing high level of consumption of these foods in underdeveloped countries, much remains unknown about the detrimental effects of phytate in legumes and cereals on mineral bioavailability in relation to human nutrition.

II. MINERALS

A. ZINC

The first direct evidence that zinc deficiency may develop in animals fed a diet composed of natural plant products (corn and soybean meal) was discovered in pigs by Tucker and Salmon.[26] They showed that zinc deficiency (characterized by depression in growth and severe skin lesions) in pigs could be cured or prevented by zinc supplementation. Later, phytate induced zinc deficiency was shown in chicks,[27,28] swine,[29] rats[30-32] and humans.[33] Animals fed animal protein-based diets containing the same level of zinc grew normally without any deficiency symptoms compared to those fed on plant protein-based diets. The phytic acid associated with plant proteins was responsible for decreased availability of zinc in diets prepared from natural plant proteins. To verify this, phytic acid was added to a casein-based diet and the growth response of chicks fed this diet and those fed soybean proteins as the only source of protein were compared.[34] Phytic acid decreased the bioavailability of zinc with the casein-based diet and produced symptoms similar to that observed among animals fed soybean protein diets containing a comparable level of phytate. O'Dell and Savage[34] attributed the decreased bioavailability of zinc to formation of a complex between zinc and phytic acid. Further, zinc forms a stable complex with phytic acid in physiological pH ranges.[5] Likuski and Forbes[35] also observed zinc deficiency symptoms in chicks after feeding them casein-based diets containing added free amino acids and phytic acid. Several studies support the view that phytic acid does decrease the bioavailability of zinc in chicks,[5,13,36] swine,[37,38] and rats.[23,39-48]

1. Animal Studies

Several bioassay methods have been used to measure zinc bioavailability in cereals, beans, and their products. Methods for assessing zinc bioavailability have included thelog dose response of weight gain, growth rate, the chelate test, the slope ratio assay of total femur zinc and weight gain, and the absorption method including both intrinsic and extrinsic labeling techniques. Zinc bioavailability values from cereals, beans, and their products relative to zinc salts are shown in Table 1. Zinc in some cereals, beans, and their products is less available either to chicks or rats than the zinc salts because of the presence of high phytate in these products. The availability of zinc also depends on the species to be used for study. For example, rats utilized less zinc from plant products than did chicks fed the same product (Table 1). Maturity of the bean also affects zinc availability to animals. By absorption method, Welch et al.[53] found that zinc in mature peas (contained 1.23% phytate) is less available to rats than that in immature peas, which had 0.17% phytate (Table 2). Further, they reported that most of zinc in mature peas is present in the form of a soluble small anionic complex (less than 1,000 MW) and which does not exist with phytate as zinc-phytate complex. In another study, Welch and House[55] observed that rats fed mature high phytate soybeans absorbed less zinc compared to immature low phytate soybeans. In contrast to this, Forbes et al.[56] found no difference in availability of zinc to rats from mature and immature soybeans.

Processing methods such as germination, milling, and fermentation appear to improve bioavailability of zinc to rats from cereals and beans.[51,52,61,62] Increased availability of zinc from germinated beans, fermented cereal products and milled fractions of cereals may be attributed to phytate hydrolysis or reduction. Franz et al.[50] observed a 30% increase in bioavailability of zinc in whole wheat bread as a result of yeast leavening (Table 1). The high bioavailability of zinc reported for white, polished rice and white flour was attributed in part to reduction in phytate content caused by milling these cereals.[62] Reinhold et al.[19] observed an increase in availability of zinc in whole wheat leavened bread as a result of phytate hydrolysis during yeast fermentation of the bread dough. A study by Ranhotra et al.[43] showed that bioavailability of zinc in soy fortified bread-based diets remained unaffected, probably because most of the phytate was hydrolyzed during bread making. Morris and Ellis[63] examined the effect of dephytinization of wheat bran on zinc utilization by rats. They concluded that phytate effect on zinc utilization can be overcome by dephytinization. In another study, Ranhotra et al.[42] evaluated cookies made with or without added phytate for zinc bioavailability in rats. Addition of phytate to cookies significantly reduced zinc bioavailability, based on the zinc content of rat femur. They related reduced zinc bioavailability from cookies made with added phytate to cookie-making process. Little or no phytate hydrolysis occured during the cookie-making process.

Dehulling and cooking of beans may not result in an increase in zinc availability.[59] Welch et al.[53] reported that cooking (autoclaving) mature peas did not affect the availability of zinc to rats. In an earlier study, Kratzer et al.[64] reported that autoclaving soybean protein increased the availability of zinc to turkey poults. They suggested that the increased zinc availability from autoclaved soybean protein is due to destruction of phytate. This was later confirmed by O'Dell,[65] who found that 88% of the phytate was destroyed in isolated soybean protein during autoclaving at 115°C for 4 h. In contrast, Lease[66] autoclaved sesame meal for 4 h at 15 psi and noticed only a small (22%) reduction in phytate content, despite the fact that there was a marked

TABLE 1
Bioavailability of Zinc from Cereals, Beans, and Their Products

Product	Assay animal	Method	Zinc bioavailability (%)	Ref.
$ZnCO_3$	Rat or chick	Growth rate	100	49
$ZnSO_4$	Rat	Slope ratio-gain	100	50
$ZnCO_3$	Rat	Slope ratio-gain	100	15
Cereals and cereal products				
Wheat	Chick	Growth rate	59	49
Wheat	Rat	Growth rate	38	49
Wheat	Rat	Chelate test	49—70	52
Whole wheat flour	Rat	Slope ratio-gain	79	50, 51
White flour	Rat	Slope ratio-gain	83	59
Corn	Chick	Growth rate	63	49
Corn	Rat	Growth rate	57	49
High lysine corn	Chick	Growth rate	65	49
High lysine corn	Rat	Growth rate	55	49
Corn	Rat	Chelate-test	40—68	52
Corn, raw	Rat	Slope ratio-gain	52	50, 51
Boiled corn	Rat	Slope ratio-gain	53	50, 51
Alkali corn	Rat	Slope ratio-gain	46	50, 51
Rice	Chick	Growth rate	62	49
Rice	Rat	Growth rate	39	49
White Rice, raw	Rat	Slope ratio-gain	99	50, 51
White Rice, boiled	Rat	Slope ratio-gain	92	50, 51
Brown Rice, raw	Rat	Slope ratio-gain	59	50, 51
Brown Rice, boiled	Rat	Slope ratio-gain	46	50, 51
Barley	Rat	Chelate test	60—75	52
Oats	Rat	Chelate test	62—74	52
Whole wheat bread (leavened)	Rat	Slope ratio-gain	104	50, 51
Whole wheat bread (unleavened)	Rat	Slope ratio-gain	70	50, 51
White bread (leavened)	Rat	Slope ratio-gain	102	50, 51
White bread (unleavened)	Rat	Slope ratio-gain	105	50, 51
Beans and bean products				
Peas, immature	Rat	Absorption	95	53
Peas, mature	Rat	Absorption	77	53
Peas, mature	Rat	Chelate test	58—80	52
Soybeans, immature	Rat	Slope ratio-gain	58	56
Soybeans, mature	Rat	Slope ratio-gain	58	56
Soybeans, immature	Rat	Absorption	89	55
Soybeans, mature	Rat	Absorption	60	55
Full fat soy flour	Rat	Slope ratio-femur	34	57
Full fat soy flour	Rat	Slope ratio-gain	54,55	57
Soybean meal, defatted	Chick	Growth rate	67	15, 49
Winged beans	Rat	Slope ratio-gain	85	58
White field beans, whole	Rat	Chelate test	48	59
White field beans, dehulled	Rat	Chelate test	48	59
Colored field beans, whole	Rat	Chelate test	75	59
Colored field beans, dehulled	Rat	Chelate test	69	59
Lupines	Rat	Chelate test	75—83	52
White beans, boiled	Rat	Slope ratio-gain	77	50
Lima beans, boiled	Rat	Slope ratio-gain	95	50
Soy beverage, freeze-dried	Rat	Slope ratio-gain	63	15
Soy protein isolate	Rat	Growth rate	44	32
Soy protein isolate (neutral form)	Rat	Slope ratio-gain	85	15
Soy protein isolate (acid form)	Rat	Slope ratio-gain	106	15

TABLE 1 (continued)
Bioavailability of Zinc from Cereals, Beans, and Their Products

Product	Assay animal	Method	Zinc bioavailability (%)	Ref.
Soy protein concentrate (neutral form)	Rat	Slope ratio-gain	66	15
Soy protein concentrate (acid form)	Rat	Slope ratio-gain	113	15
Soy protein concentrate (made by 60% ethanol leaching)	Rat	Slope ratio-grain	67	242
Soy protein concentrate (made by hot water leaching)	Rat	Slope-ratio-grain	81	242
Calcium-tofu	Rat	Slope ratio-gain	51	54
Magnesium-tofu	Rat	Slope ratio-gain	51	54
Soy-based infant formula	Rat	Slope ratio-femur	67	60

Note: Zinc bioavailability values are relative to inorganic salt = 100.

TABLE 2
Effect of Seed Maturity on Bioavailability of Zinc in Peas[53]

Product	Assay animal	Bioavailability (%)
Peas, immature, autoclaved	Rat	94.9
Peas, mature, autoclaved	Rat	77.4
$ZnSO_4$	Rat	88.3

increase in zinc availability to chicks. Obviously, there is a difference between soybean protein and sesame meal in responding to heat treatment. The mechanism of how zinc is held by phytic acid is still unknown. Zinc may be held by one or several bonds to the phytic acid and breaking of one of these bonds by autoclaving would allow it to become available to the chick. It is also likely that zinc may be held by something other than the phytic acid which was destroyed by autoclaving and thus released zinc to be used in the presence of phytate.[66]

Lease[67] evaluated the bioavailability of zinc from zinc-phytate complexes isolated from oilseed meals in chicks by an *in vitro* digestion method. She showed that the zinc of sesame meals and safflower was present in an insoluble, nondialyzable Ca-Mg-Zn-phytate complex at the intestinal pH and was poorly available to experimental chicks. Forbes and Parker[57] have shown that zinc added to rat diets in the form of full-fat whole soy flour was significantly less biologically available than zinc added as zinc carbonate to an egg white protein diet. Bioavailability of zinc also varies from product to product.[68] For example, zinc is better utilized from a full-fat soy flour than from a soy concentrate or soy isolate.[17,69-72] Several soybean products such as full-fat soy flour, soy beverage (freeze-dried), soy protein isolate (neutral form), soy protein isolate (acid form), soy protein concentrate (neutral form), soy protein concentrate (acid form), and a commercial soy protein concentrate were evaluated for bioavailability of zinc to rats using slope ratio technique.[15,57,71,73,74] In soybean products, zinc bioavailability varied from product to product (see Table 1). The bioavailability of zinc from soybean products is low compared to the availability of zinc from zinc carbonate. The acid forms of the isolates and concentrates showed excellent bioavailability for zinc relative to neutralized products (isolates and concentrates) prepared under identical conditions.[15,72,74] The differences in bioavailability of zinc from soy

TABLE 3
Effect of Dietary Calcium Level on Weight Gain[15,54]

Diet	Dietary calcium level (%)		
	0.4	0.7	1.2
Egg	117	105	95
Ca-tofu	112	75	46
Mg-tofu	114	70	46

Note: Rats were fed 9 ppm zinc from calcium- or magnesium-precipitated tofu or from zinc carbonate added to egg white diets.

products are most likely due to the use of various processing conditions employed during their preparation.[15,73] Erdman et al.[15] believe that the stable protein-phytate-zinc complexes formed during processing are responsible for the reduced mineral absorption in soy products rather than specific phytate content in these products. More research is needed to identify the processing conditions and steps that affect formation of protein-phytate-zinc complexes.

Calcium also accentuates the effect of phytate on zinc availability.[75,246] The interrelationship among calcium, phytate, and zinc has been studied in different species of animals.[13,40,76-79] Calcium aggravates zinc deficiency when it is added to diets based on plant products that are high in phytate or diets with added phytic acid. Calcium and zinc appear to have a synergistic effect in the precipitation of phytate. To determine the mechanism whereby these two minerals interact with phytic acid, Byrd and Matrone[80] studied the influence of calcium on incorporation of zinc into the phytate complex. They found that when the molar ratio of the complexing of zinc was 1:1 or 2:1, calcium decreased the complexing of zinc to phytate; when the ratio of calcium to zinc was 100:1, the presence of excess calcium in the solution increased incorporation of zinc to the point where 99% of the zinc was present as a phytate complex. A similar effect was also found by Oberleas et al.[39] It is understandable that when zinc and calcium are present at high levels, calcium competes for positions on the phytate molecule, thereby reducing the amount of zinc precipitate. However, under practical feeding conditions those rations associated with the development of parakeratosis usually contained 30 to 100 ppm zinc and 1 or 2% calcium. The presence of excess calcium may provide an explanation for zinc deficiency symptoms observed in the above case. It was postulated that calcium increases the total cationic environment sufficiently to initiate a coprecipitation with zinc to form Zn-Ca-phytate complex. This resultant complex is reported to be less soluble than Zn-phytate at pH 6.0., the approximate pH of the upper intestine, where most of the absorption of these divalent cations occurs. Therefore, zinc in the presence of excess calcium would not be as available for intestinal absorption and aggravates the zinc deficiency. The sparingly soluble calcium phytate complex has been reported to have little effect on zinc availability and only the soluble phytate in the diet has shown to have the binding effect.[13] Forbes et al.[54] compared the bioavailability to rats of zinc contained in calcium sulfate and magnesium chloride – precipitated soybean tofu to that of zinc added as the carbonate to egg white diets using slope ratio technique. Ca-tofu and Mg-tofu had about 0.5% phytate phosphorus (1.8% phytate). The total dietary calcium level in all diets adjusted to 0.7% with calcium carbonate. The relative bioavailability of zinc from two tofu preparations was 51% as measured by slope ratio weight gain and use of two common coagulants (calcium sulfate and magnesium chloride) in preparation of soybean tofu did not influence zinc bioavailability. In a separate experiment, Forbes et al.[54] evaluated the effect of three dietary calcium levels on the response of rats consuming a phytate-free diet and diets containing tofu. Zinc was supplied at 9 ppm in all diets. The response of tofu-fed rats compared to zinc carbonate-fed rats appeared to be quite similar at 0.4% dietary calcium (Table 3). However, performance of rats reduced as dietary calcium level increased to 1.2%. They attributed the inhibitory effect of calcium on zinc bioavailability in high phytate diets to formation of insoluble complexes (calcium-zinc-phytate) in the gastrointestinal tract as a result of presence of sufficient calcium in diets. They further indicated that these complexes are not formed during the tofu preparation process. Magnesium also accentuates the effect of phytate on zinc bioavailability.[79] Forbes et al.[79] found that magnesium exerts a less pronounced effect on zinc utilization in phytate containing diets than does calcium.

Addition of EDTA (ethylenediaminetetraacetate), a chelating agent, to the diet neutralizes the deleterious effect of soluble phytate on the zinc availability and increases zinc absorption both in chicks[13] and rats.[39] The mechanism by which EDTA makes zinc more available to animals is not clearly known. It has been suggested that EDTA competes with phytate in chelating zinc to form a Zn-EDTA complex and thus interferes with the formation of Ca-Zn-phytate complex.[6] The soluble Zn-EDTA complex can either be absorbed as such or can release zinc ion to the intestinal absorption and thus counteract the detrimental effect of the phytate.

Several studies have been reported about the possible role of fiber in the mineral bioavailability in cereals and legumes.[18,20,21,81] These investigators suggest that dietary fiber may contribute towards reduced mineral bioavailability, along with phytate, from cereals and legumes. Such synergistic effects of fiber were investigated by Reinhold et al.[14] who found that the unleavened whole wheat bread containing high fiber had a more detrimental effect on zinc balance than sodium phytate supplied daily at the same level as that in the bread. In contrast to these observations, several other studies with experimental animals suggest that fiber does not contribute to reduced zinc utilization.[82] In one study,[22] rats were fed diets containing either wheat bran (having both phytate and fiber), extracted bran fiber, or phytate. Rats that received bran or phytate showed impaired zinc utilization as assessed from decreased growth rates, whereas the rats receiving fiber grew at the same rate as those receiving control diets containing no supplements. Based on these observations, Davies et al.[22] concluded that phytate is the major determinant in the reduction of zinc availability to rats rather than fiber in the wheat bran. In a subsequent study, Davies[83] came to similar conclusions, while using dietary supplement of individual components of dietary fiber namely cellulose, hemicellulose, and pectin to rat diets. Weingartner et al.[84] also showed that inclusion of soybean hulls to soy flour-based diets had no significant effect on the bioavailability of native zinc or of added calcium to rats. They suggest that soybean hull fiber plays no role in reducing mineral absorption when fed at levels normally found in whole soybean products.

2. Human Studies

Zinc deficiency in humans was first reported by Prasad et al.[33] In this study, Egyptian boys consumed a diet consisting mainly of phytate-rich cereals and beans. The patients, who were characterized by dwarfism and hypogonadism, showed an improvement in growth and sexual maturity in response to zinc supplementation of a hospital diet. Analysis of foods consumed by humans suffering from zinc deficiency symptoms revealed that the amounts of zinc contained therein were adequate according to published nutrient requirement literature. Later it was realized that the presence of phytate in plant products was an important factor in the reduction of zinc absorption from foodstuffs.[85] The effects of phytate on zinc deficiency in humans were strengthened by the reports from Iran.[14,86-91] In these studies, zinc deficiency occurred when unleavened flat bread was consumed in greater amounts than leavened bread. The flat breads are prepared from high-extraction wheat meal and contained higher amounts of phytate than the leavened bread (Table 4). People consume very little animal protein in these countries.

Based on U.K. food composition tables, Davies[82] assessed phytate intakes in the U.K. He reported that phytate consumption in the U.K. is in the range of 600 to 800 mg daily. About 70% of the phytate intake is derived from cereal products, 20% from fruit, and the remainder from vegetables and nuts. In developing countries, where many households consume diets primarily

TABLE 4
Phytate Concentrations of Iranian Flat Breads[89,90]

Bread	Flour extraction (%)	Phytate (mg/100 g)	Treatment
Bazari	75	326	Leavened
Sangak	85—90	388	Leavened
Tanok	90—100	684	Unleavened

TABLE 5
Apparent Zinc Absorption in Young Men[93]

Diets	Mean zinc absorption[a] (%)
Basal diet	34.0 ± 6.2
Basal diet + alpha-cellulose	33.8 ± 2.9
Basal diet + phytate[b]	17.5 ± 2.5

[a] Mean ^{67}Zn absorbed
[b] Sodium phytate 2.34 g added to basal diet.

Note: Basal semipurified liquid diet comprised of egg albumin as protein source, dextrimaltose, corn starch, sucrose, cottonseed oil, anhydrous butterfat, macrominerals, trace elements, and vitamins.

consisting of mixtures of cereals and legumes, the average daily phytate intake may be higher when compared to developed countries, where the diets contain primarily animal proteins. Within the U.S., adolescents are the population most at risk relative to zinc deficiency. The diet of this group includes cereals, peanut butter, and pasta as a major component. The other at risk group is vegetarians, or those substituting vegetable protein-based synthetic meats for animal protein.[92]

Several researchers investigated zinc absorption in humans from composite meals containing varied levels of phytate.[93-97] Most of these studies concluded that high levels of phytate in the meals decreases zinc absorption. Turnlund et al.[93] studied the effects of alpha-cellulose and phytate on zinc absorption in young men. Zinc absorption was measured by monitoring fecal excretion of a stable isotope of zinc given in a semipurified liquid formula diet supplying 15 mg of zinc daily. Average zinc absorption was 34.0% for the basal diet, but fell to 17.5% when 2.34 g of phytate as sodium phytate was added to basal semipurified liquid formula (Table 5). They[93] concluded that phytate inhibits zinc absorption and high levels of dietary phytate could result in zinc deficiency in man. However, it is not known whether or not the effect would be the same if the same level of phytate were consumed in a natural food diet. Navert et al.[94] obtained a substantial improvement in zinc absorption from 9.6% to 19.8% by prolonged (16 h) leavening of bread containing moderate amounts of bran and phytate. In another study, Sandstrom et al.[96] found marginal effects on zinc absorption in men when meat protein was partially replaced with soy protein in composite meals.

Possible effects of fiber on zinc availability from foods in humans were reported. Reinhold et al.[18] demonstrated the effects of fiber on zinc, phosphorus, and calcium balances in humans. Subjects were maintained on a low fiber diet, which provided 50% of energy intake. After 14 days they consumed the same diet but in addition they received 10 g of cellulose dispersed in 150 g apple compote. Fecal excretion and balance of zinc, calcium, and phosphorus were monitored throughout the experiment. They found negative zinc balances in humans who consumed lower fiber diet and additional cellulose. Several other investigators also reported decrease of zinc absorption in humans due to increased content of bran or dietary fiber in meals.[94,95,98-100] Sandstrom et al.[95] found that percentage absorption of zinc was less (17%) from a whole wheat bread meal than from a white bread meal (38%).

There are other studies which failed to demonstrate impaired zinc utilization due to dietary fiber in humans. Sandstead et al.[101] tested five sources of dietary fiber, including wheat bran and concluded that fiber did not have any effect on zinc balance in adult men. Guthrie and Robinson[102] conducted zinc balances in four young women, with and without daily supplements of wheat bran (14 g/day), and found no overall difference in zinc metabolism. Morris and Ellis[103] compared zinc balance of adult men consuming whole or dephytinized wheat bran muffins. The intakes of dietary fiber and zinc were the same for each diet. They found positive zinc balance in all subjects regardless of the type of bran muffins consumed. The mean zinc balance was 0.7 mg per day higher in subjects eating whole bran muffins than those eating dephytinized bran muffins. Several factors may account for these conflicting results. Differences in sources of fiber and its composition, and/or dietary calcium, zinc, phytate contents, and protein source and content contribute to varied findings.[82] The zinc availability to humans from fiber and phytate-rich whole meal bread can be improved by increased intakes of animal protein.[95]

3. Phytate/Zinc Molar Ratio for Predicting Zinc Bioavailability

Use of phytate/zinc molar ratio to predict dietary zinc bioavailability from phytate-rich foods was first suggested by Oberleas.[104] Applicability of phytate/zinc molar ratio as a predictor of zinc bioavailability to rats has been investigated using sodium phytate in semipurified diets,[24,78,105] soybeans and soy products,[15,25,74,106] cereals and legumes,[50,61,107] wheat bran[63] and seven ready to eat and two quick-cooking breakfast cereals.[119] Phytate/zinc molar ratios of greater than 12 to 15 are reported to be associated with chemical zinc deficiency. Most cereals, legumes, and their products contain higher (more than 15) phytate/zinc molar ratios (Table 6) and zinc may not be readily available for absorption in animals and humans from these foods and food products when consumed. However, bioavailability of zinc from phytate-rich foods and food products can be increased either by supplementation of zinc or enzymatic reduction of phytate to lower phytate/zinc molar ratio to less than 10.[105,106] Harland and Harland[120] decreased phytate/zinc molar ratios to more favorable levels in rye, white, and whole wheat breads made with addition of yeast and increase of rising times. With these two treatments, they decreased phytate/zinc molar ratio from 39 to 10 in rye bread, 6 to 4 in white bread, and 24 to 16 in whole wheat bread.

Davies and Olpin,[24] Morris and Ellis[63] and Erdman et al.[15] indicated the importance of calcium content of the diet to the phytate/zinc molar ratio on zinc bioavailability. Presence of higher level of calcium decreases zinc bioavailability from phytate-rich foods. Moderate levels of calcium intake in a diet containing phytate/zinc molar ratio less than 10 provides adequate available zinc. Morris and Ellis[63] reported that dietary phytate/zinc molar ratios of 12 or less did not depress growth of rats if the diets contained a moderate level (0.75%) of calcium. Addition of 1% calcium to the diets enhanced the effect of phytate on zinc utilization. Recently, Erdman et al.[15] concluded that phytate/zinc molar ratios alone do not predict zinc bioavailability from soy foods, but food processing conditions and several other factors play a role in the bioavailability of zinc from these foods. This was clearly demonstrated in a study by Erdman et al.[74] They found significant differences in bioavailability of zinc from soy isolate or concentrates with similar composition and phytate/zinc molar ratios (Table 7). They prepared two soy isolates or concentrates by identical procedures using acid precipitation method. One acid-precipitated product was freeze-dried without neutralization and the other was neutralized prior to freeze-dehydration. These two soy products (isolates or concentrates) were then fed to rats to determine the zinc bioavailability. Zinc from acid-precipitated isolate or concentrate (without neutralization) produced excellent growth in rats, whereas rats fed acid-precipitated isolate or concentrate

TABLE 6
Phytic Acid/Zinc Molar Ratios of Beans, Cereals, and Their Products[15,50,54,56,58,61,63,72,81,103,106,108-120]

Product	Phytic acid/zinc molar ratio
Beans	
Green immature soybeans	10.5 —18.8
Soybeans	26.0 —44.0
Winged beans (whole)	21.7—30.4
Lima beans	20.0—79.0
Navy beans	22.0
Navy beans (dehulled)	51.2
White beans	33.0
Great Northern beans	72.4
Pinto beans (dehulled)	43.6
Black beans (dehulled)	49.7
Peas, mature	24.7—32.9
Pea cotyledons	25.9
Pea germ	16.7
Black gram (whole)	26.0—26.7
Black gram (dehulled)	44.0
Chick peas (whole)	17.3
Chick peas (dehulled)	24.7
Green gram (whole)	26.5
Red gram	40.8
Kidney beans	31.6
Lentils (whole)	15.9
Lentils (dehulled)	14.4
Blackeye bean	35.7
Bean products	
Full-fat soy flour	28.0—29.0
Soy isolate	71.8
Soy concentrate	44.0
Soy beverage (freeze-dried)	26.0
Low phytate soy concentrate	13.0
Autoclaved soybeans	30.0
Ca-tofu	27.6—32.0
Mg-tofu	26.1—29.0
Peanuts, roasted	7.0
Peanut butter	42.0
Lima beans, boiled	20.0
Navy beans, boiled and drained	34.0
White beans, boiled	31.0
Chick peas, whole, cooked	13.9
Chick peas, dehulled, cooked	20.7
Chick peas, boiled and drained	14.0
Peas, autoclaved	38.8
Peas, blanched	34.9—41.8
Germinated, freeze-dried peas (8 days germinated)	23.0—25.4
Green peas, immature, canned and drained	3.0
Black gram, whole, cooked	22.7
Green gram, whole, cooked	23.8
Red gram, cooked	23.5
Kidney beans, cooked	26.6
Lentils, whole, cooked	8.5
Lentils, dehulled, cooked	7.2
Blackeye bean, cooked	26.1
Great Northern bean major density fraction	84.8
Navy bean, high protein flour	57.0

TABLE 6 (continued)
Phytic Acid/Zinc Molar Ratios of Beans, Cereals, and Their Products[15,50,54,56,58,61,63,72,81,103,106,108-120]

Product	Phytic acid/zinc molar ratio
Pinto bean, high protein flour	46.1
Black bean, high protein flour	51.5
Idli, black gram and rice fermented food	23.1
Papadum	36.7
Cereals and cereal fractions	
Wheat	28.7
Whole wheat flour	25.0—34.0
Wheat germ	28.0
Hard wheat bran	40.0—44.0
Soft wheat bran	46.0—52.0
All purpose wheat flour	40.0
Enriched white flour	5.0
Rye	26.3
Triticale (6X)	20.7
Triticale (8X)	25.8
Oats	30.3
Corn	33.3
Brown rice	47.0
White rice	11.0—17.0
Cereal products	
Brown rice, boiled	38.0
White rice, boiled	4.0
Hard wheat bran, autoclaved	40.0
Ready to eat wheat cereal	62.0
Shredded wheat cereal	52.0
Whole wheat bran muffins	10.5—13.6
Enriched white bread	6.0—11.0
Whole wheat bread	21.0—24.0
Whole wheat bread, unleavened	16.0
Chapati, unleavened flat bread	18.1—28.1
Corn, boiled	30.0
Corn chips	41.0
Corn flakes	16.0
Pop corn, popped plain	15.0
Ready to eat cereal flakes	16.0
Rolled oats, cooked	22.0
Saltine crackers	33.0
Regular cooked farina	7.0
Macaroni, cooked tender stage	16.0
Granola	29.0
Rye bread	38.0—58.0
Middle eastern breads	
Bazari (leavened)	10.8
Sangak (leavened)	12.8
Tanok (unleavened)	22.6
Italian natural foods	
Brans	37.5—39.8
Breads	2.9—11.6
"Grissini"	18.7—32.6
Crackers	11.6—19.6
Rusks	14.1—16.0
Biscuits	11.3—25.8

TABLE 7
Composition of Soy Products and Their Relative Bioavailability of Zinc to Rats[15,74]

	Soy concentrates		Soy isolates	
Constituent	Acid form	Neutral form	Acid form	Neutral form
Solids (%)	93.6	92.8	98.9	98.0
Protein (%)	80.8	80.3	94.1	92.3
Ether extract (%)	0.2	0.3	0.2	0.2
Zinc (ppm)	30.0	34.0	43.0	45.0
Phytic acid (%)	1.98	1.93	1.52	1.52
Phytate/zinc molar ratio	66.0	57.0	35.0	34.0
Relative bioavailability of zinc (%)[a]				
Weight gain	113.0	66.0	106.0	85.0
Leg bone zinc	48.0	29.0	64.0	46.0

[a] Data points represent comparisons of slopes of responses of test diets (containing soy) with responses of control diets (without soy products) with added minerals as the carbonate × 100.

(neutralized) had significantly reduced growth responses (Table 7). The differences in growth response of rats to soy isolates or concentrates may be attributed to the processing conditions employed during their preparation from soybeans. Further, the difference may also be due to formation of stable protein-phytate-zinc complexes in the freeze-dried neutral product (soy isolate or concentrate).[74] During drying of soy isolates or concentrates, the protein-phytate-zinc mineral complexes form tightly bound complexes. The exclusion of water from these complexes could lead to thermodynamically stable complexes that are resistant to complete proteolytic digestion in the gastrointestinal tract. Thus, the zinc and phytate complexed to protein would be poorly absorbed.[15] In addition to phytate/zinc molar ratio and calcium content, several other factors such as methods of food processing and processing conditions used during preparation of a food must also be considered when predicting zinc bioavailability from foods.

Results from studies by Greger et al.[121] and Young and Janghorbani[122] indicated that zinc bioavailability in healthy humans was not affected by a soybean based diet (with phytate/zinc molar ratio 5). However, Greger et al.[121] reported that zinc retention was affected when healthy human subjects received diets containing phytate/zinc molar ratio of 9. Navert et al.[94] studied zinc absorption in man from bread meals containing 10 g of wheat bran fermented for varying lengths of time. They reported that zinc absorption in man increased from 9.6% to 19.8% when the molar ratio of phytate/zinc decreased from 17 to 4 in bread meals by 16 h of fermentation. Recently, Morris and Ellis[103] conducted a human metabolic study to test the concept of dietary phytate/zinc molar ratio as a predictor of zinc bioavailability to humans. They attained phytate/zinc molar ratios of about 1 and 12 in foods by incorporating 36 g/day of whole or dephytinized wheat bran. Foods used daily in this study were those commonly consumed in the U.S. The mean daily intake of zinc was 17 mg and of neutral detergent fiber 16 g. They[103] found no difference in human zinc balance i.e., the mean zinc balance was 2.74 mg per day when dietary phytate/zinc molar ratio was about 12, and 2.0 mg per day when the ratio was about 1. In a subsequent study, Morris et al.[123] observed slightly negative zinc balances in adult men when they received daily zinc intakes of 10 mg in a diet containing high phytate and phytate/zinc molar ratio of 27. It appears that a diet containing a phytate/zinc molar ratio of 10 to 12 is not going to affect zinc utilization.

According to Ellis et al.,[124] the majority of the U.S. population consumes a diet that has a phytate/zinc molar ratio of less than 10, which provides less than the recommended dietary allowance of zinc. Vegetarians, particularly those who eat primarily cereal and bean products, likely have a diet that has phytate/zinc molar ratios exceeding the value of 10. Humans consume a variety of foods that contain a wide range of phytate/zinc molar ratios.[125,126] Oberleas and Harland[81] suggest that the phytate/zinc molar ratios can be used to estimate relative risk of having an adequate intake of zinc in a diet and in planning menus to select the combination of foods that supply most available zinc to daily diet.

Based on the influence of calcium and phytate on zinc bioavailability, Davies et al.[127] reported on a dietary ratio, (phytate) (calcium)/(zinc) (moles/kg) for predicting zinc bioavailability. They tested this ratio in rats and found that ratios exceeding 3.5 (moles/kg) may result in a zinc responsive growth impairment. The validity of this ratio for predicting the zinc availability in humans is not established. Fordyce et al.[128] investigated this ratio as a predictor of zinc bioavailability from several processed soybean products. They concluded that (phytate) (calcium)/(zinc) ratio can be used as a predictor of zinc bioavailability for similar processed products but not for products prepared by using different process techniques. The adaptability of this ratio for predicting zinc availability to man and animals from different foods containing phytate and calcium remains to be studied. Some of the uncooked and cooked beans, unleavened flat breads, and Asian lacto-ovovegetarian diets contain high phytate × calcium/zinc molar ratios.[129,130] The availability of zinc may be poor from these foods if one considers the ratio of (phytate) (calcium)/(zinc).

B. IRON

Heme iron in animal products has been reported to be more readily available for absorption than that in plant products[131] because plant products contain a number of inhibitors, including phytic acid. Fitch et al.[132] observed a lower absorption of iron in rhesus monkeys fed soybean protein than those fed casein. A few studies[23,133,134] have demonstrated inhibitory effects of phytate on iron absorption by adding sodium phytate to diets. However, Hunter[135] concluded that high levels of sodium phytate in the diet of rats does not inhibit utilization of dietary iron. The conflicting results on effects of phytate on iron absorption may be due to differences in experimental conditions and presence of inhibitory components other than phytate in the diets.

1. Animal Studies

Several researchers[23,136,137] reported that addition of phytate to foods or the presence of high phytate in foods substantially reduces the bioavailability of dietary iron. Others have found that feeding relatively high levels of phytate had little or no adverse effect on iron availability and absorption by animals.[8,139–144,243] Much of this controversy may be attributed to differing types of experimental animals, techniques used by these investigators, and the presence of intestinal phytase that might liberate iron as the phytate is degraded. Patwardhan,[145] Pileggi,[146] Roberts and Yudkin,[147] and Ranhotra et al.[142] showed that rats possess an intestinal phytase. Its quantitative activity has not been thoroughly investigated. The possible presence of phytase has been suggested as the reason why rats can utilize iron from phytate rich cereals, legumes, and other food products (Table 8).

Welch and Campen[150] reported that intrinsically labeled iron (^{59}Fe) in mature soybeans is more available than iron in immature soybeans, even though mature beans contained approximately three times more phytic acid (Table 9). They concluded that apparently immature beans contain a factor or factors other than phytic acid which impair iron availability. Based on studies with rats, iron availability from soybeans and its protein products has been reported to range from 28.5 to 92.0% (of inorganic iron).[148,150-153] Steinke and Hopkins[152] compared hemoglobin repletion in rats fed one of three soy protein isolates to rats fed ferrous sulfate. They obtained a mean relative iron bioavailability of 61% from soy protein isolates and further concluded that inorganic iron added to diets containing isolated soybean protein had bioavailabilities similar to that of iron present in the soybean. In a separate experiment, Schricker et al.[153] found no significant differences in absorption of iron to rats from soy flour, soy protein concentrate, and soy protein isolate. Bioavailability (determined from whole-body retention curves of ^{59}Fe in rats)

TABLE 8
Relative Bioavailability of Iron to Rats From Cereal and Legume Products

Product	Relative bioavailability (%)	Ref.
Ferrous sulfate	100	148
Ferrous ammonium sulfate	100	58
Hard wheat		
Whole grain	90	149
Bran	86	149
Germ	92	149
Shorts	92	149
Soft wheat		
Whole grain	88	149
Bran	98	149
Germ	91	149
Whole wheat breads	53—115	149
Wheat cereals		
Ready to eat whole wheat	100	149
Ready to eat bran	89	149
Instant whole wheat	72	149
Winged bean	89	58
Soybean flour	81	148
Soy protein concentrate	92	148
Soy beverage (freeze-dried)	66	148

Note: The method used in all cases was the slope ratio.

TABLE 9
Absorption of ^{59}Fe From ^{59}Fe-Labeled Soybean Seeds in Iron-Depleted Rats[150]

Treatment	Phytic acid (% dry weight)	Iron absorption (% of dose)
Immature soybeans	0.65	30.2
Mature soybeans	1.77	55.5
^{59}FeCl$_3$	—	50.7

of iron from defatted soybean flour and its fractions (soybean hulls, soy protein isolate, and soy whey) has been evaluated.[154,155] Absorption of iron from soy whey (supernatant at pH 4.0 contains mainly soluble proteins and carbohydrates) was higher than from soy protein isolate, soybean hulls, and defatted soybean flour, but the retention values for iron were not significantly different among soybean fractions.[155] Peas and lima beans were also found to be better sources of bioavailable iron than navy beans.[156] However, the chemical nature of iron in beans is not known. In soybeans, a portion of iron may be present in the form of monoferric phytate.[106]

Processing may contribute to an increase in the availability of iron from beans. Autoclaving (108.4°C for 30 min) of isolated soy protein improves iron bioavailability.[152] Using the chick hemoglobin repletion method, Rodriguez et al.[157] studied the effect of heat and partial removal of phytate on *in vivo* availability of iron from soy protein. Heating of soy protein at 120°C for 20 min and removal of 75% of the phytate from soy protein increased iron bioavailability by 65 to 77% and 6 to 11% respectively. They attributed the increases in bioavailability of iron to digestion of protein and protein-iron-phytate complexes and subsequent release of endogenous iron.

Morris and Ellis[149,158] isolated and characterized a major iron component from hard wheat bran as monoferric phytate (Figure 1) and showed that over 60% of iron in wheat bran is present

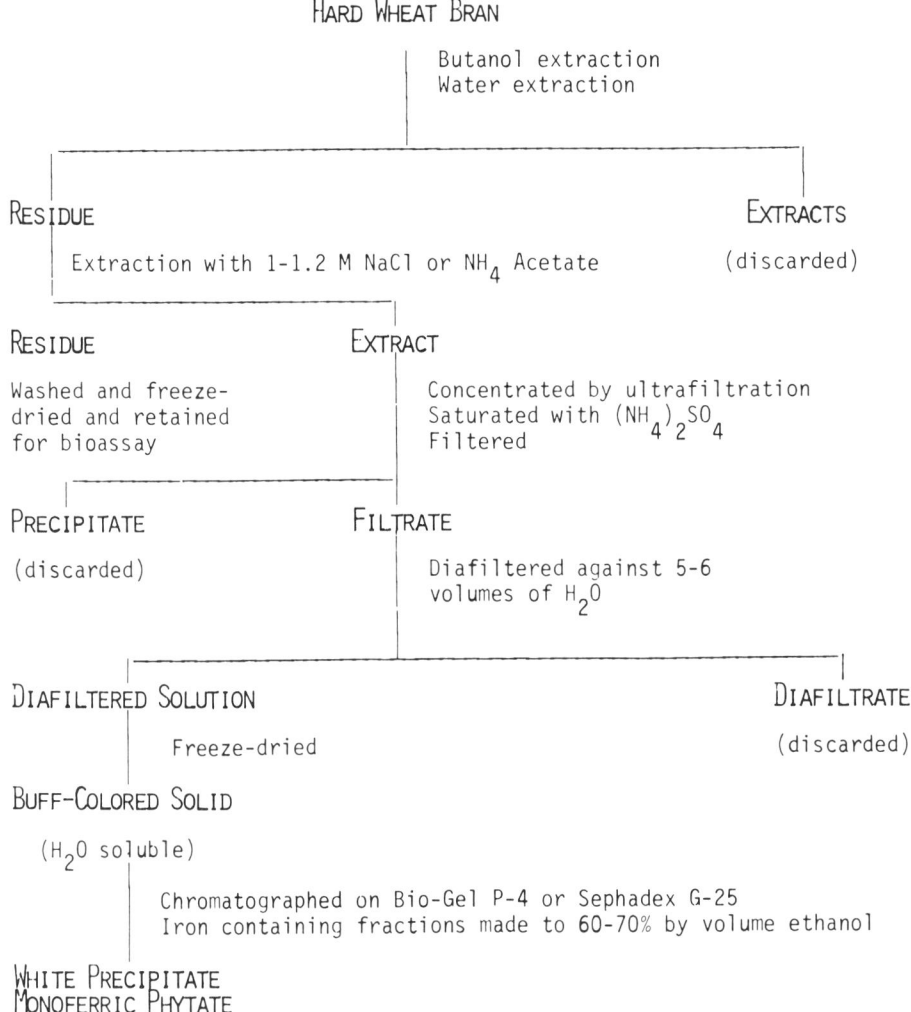

FIGURE 1. Isolation and purification of monoferric phytate from wheat bran. (From Morris, E. R. and Ellis, R., *Baker's Digest*, 50, 28, 1976. With permission.)

as monoferric phytate. May et al.[159] reported that isolated monoferric phytate had identical Mossbauer spectral parameters (quadruple splitting and isomer shift) to that of wheat bran and whole wheat grain. About 99% of the iron from monoferric phytate was found to be available for absorption by rats. Morris and Ellis[160] used rats to test the bioavailability of monoferric phytate fed along with bread or baked in bread. They found no differences in the relative bioavailability values (calculated by slope ratio) of the reference salt fed with bread (100) or baked into bread (96), or of monoferric phytate either baked in bread (101) or fed with bread (99). Several studies[149,158,161-163] demonstrated that monoferric phytate is a good source of iron and highly available to rats and dogs. Morris and Ellis[149] evaluated wheat and its milling fractions, whole wheat breads and wheat cereals for bioavailability of iron to the rat. The relative bioavailability values for iron in wheat and its milling fractions (shorts, germ, and bran) ranged from 86 to 92% with an average of about 90%, based on response to ferrous ammonium sulfate (Table 8). In commercially available whole wheat breads, the relative bioavailability values for iron varied from 53 to 115%. Instant cereal had a low relative iron bioavailability value compared to ready to eat whole wheat and bran (Table 8).

2. Human Studies

Three decades ago it was reported that phytate impaired iron absorption in humans.[138,164-167] Widdowson and McCance[164] found mean iron balances of 1.6 and 0.25 mg/d for individuals consuming white and brown bread diets, respectively. Brown bread was made from 92% extraction flour and contained 4 times as much phytate P as did the white bread. They attributed the lower iron balances in individuals consuming brown bread to its higher phytate P content. In a study with adolescent boys, Sharpe et al.[167] found that addition of sodium phytate to milk (0.2g/200ml) decreased iron absorption by 15-fold. However, the same quantity of phytate in the form of oatmeal had much less effect. Hussain and Patwardhan[168] reported a decrease in the iron balance in humans consuming a Indian diet (paratha) in which the percentage of phytate P to total P was increased from 8 to 40 by addition of sodium phytate. Based on iron utilized for red blood cell production, Turnbull et al.[169] demonstrated that the addition of sodium phytate to 5 mg of ferrous ascorbate resulted in a 50% reduction of iron absorption in humans. Addition of phytate to foods or the presence of high concentrations of phytate in foods has been reported to have an influence on bioavailability of iron.[170-174] The bioavailability of iron to humans is poor from plant foods (wheat germ, butter beans, and brown and green lentils) that contain large amounts of phosphorus in the form of phytate as compared to vegetables containing ascorbic acid.[175] Lynch et al.[176] determined iron bioavailability in a variety of legumes to humans by radioisotopic measurements of iron absorption. Soybeans, black beans, lentils, mung beans, and split peas were prepared as soups and served to individuals for measuring iron absorption. Based on the iron absorption values (ranged from 0.84 to 1.91% for legumes), they concluded that legumes are poor sources of dietary iron because of presence of several inhibitory factors such as tannins and phytate. Relatively low concentrations of tannins in legumes appear to contribute to the lowering of iron absorption.[177] In contrast, Young and Janghorbani[122] reported that iron in soy protein isolates is readily available for absorption by humans. The results of Beard et al.[243] also suggest that phytate in soybeans and soy protein products has no effect on iron absorption in humans.

Several other researchers have studied the effect of wheat bran on iron absorption by individuals.[99,100,160,178-181] Bjorn-Rasmussen[179] found a direct negative relationship between nonheme iron absorption and the amount of wheat bran added to bread. The inhibition of iron absorption by wheat bran is attributed to the phytate content of the bran. Anderson et al.,[178] however, found no change in iron balance for individuals who consumed breads made with increased amounts of bran. There may be components other than phytate present in wheat bran that are responsible for low iron absorption. Simpson et al.[180] investigated the mechanisms for the inhibiting effect of wheat bran and the role of phytate from wheat bran on iron absorption by humans. The iron absorption studies were done with muffins containing either dephytinized bran or untreated bran. Individuals were served meals consisting of a milk shake with 100 mg ascorbic acid and 2 bran or plain muffins. Absorption of nonheme iron was measured by the intrinsic tag method. They concluded that iron absorption was greater with dephytinized bran muffins compared to untreated bran or plain muffins. Dephytinized bran inhibited absorption of nonheme iron to almost the same degree as the untreated bran. Other results in the study of Simpson et al.[180] indicated that inhibition was not caused by the phytate present in wheat bran but rather by a water soluble phosphate-rich fraction.

In a separate study, Morris and Ellis[160] compared iron balance of humans who consumed untreated or dephytinized bran continuously over several weeks. The mean daily iron balance was 0.9 mg greater for subjects who consumed untreated bran compared to those who consumed dephytinized bran. Iron balances were not increased in the subjects who consumed dephytinized bran. They[160] indicated that the effect of wheat bran cannot be unequivocally attributed to either phytate or fiber alone but the interactions between fiber, phytate, and iron may be responsible for the effect of wheat bran on iron absorption. Recently, Morris et al.[123] completed a second iron balance study in which 3 levels of phytate were fed to humans. They concluded that the iron balances were predominately negative regardless of dietary phytate level. Monoferric phytate is the major fraction of iron in wheat and wheat bran.[159] Simpson et al.[180] measured absorption by humans of free monoferric phytate from meals of both high and low bioavailability. Absorption was measured by the extrinsic tag method. They found that monoferric phytate is absorbed by humans to the same extent as an extrinsic radio iron tag.

Processing also influences iron availability from food products. Cook et al.[182] reported differences in nonheme absorption by humans fed meals containing full-fat soy flour, texturized soy flour and isolated soy protein. Iron bioavailability was lower for soy protein isolate than soy flour. Baking improves bioavailability of iron in food products such as soy protein isolate and whole soybeans. Limited information is available on the bioavailability of iron in many legumes and cereals except soybeans and wheat. In general, iron is poorly absorbed from cereals and legumes, with the exception of wheat.[183]

C. CALCIUM

Phytic acid in foods may interact with calcium and cause a reduction in its bioavailability to humans and animals. The interest in phytate in nutrition began with the demonstration by Mellanby[184] that certain cereals were anticalcifying and rachitogenic for puppies. He found that more severe rickets in puppies developed when the diet consisted mainly of oatmeal, maize, or whole wheat flour. Subsequently, Mellanby[185] reported that the rachitogenic agent in cereals could be destroyed or eliminated by either boiling with 1% HCl or seed germination. The compound responsible for such effects was later identified as phytic acid.[10,186] Mellanby[187] induced rickets in puppies raised on a low calcium diet containing added phytate, whereas controls raised on the same diet without added phytate did not become rachitic.

1. Animal Studies

High phytate intakes reduce calcium absorption and utilization of calcium in bone formation in animals.[203,204] Few studies have been reported on the bioavailability of calcium in legumes and cereals. Nwokolo and Bragg[36] showed that for growing chicks the availability of calcium in protein supplements (soybean meal, cottonseed meal, rapeseed meal, and palm kernel meal) is adversely affected by phytate.

Erdman[3] reported that rats derive only 10% of their calcium requirement from the soy when diets contain 40% soy flour. Erdman et al.[70] compared the bioavailability of calcium added to three soy products (full-fat soy flour, freeze-dried soy beverage, and a commercial soy concentrate), with calcium added to casein-based diets. They found that calcium is well utilized when added to any of the soy products. In a separate study, Forbes et al.[71] demonstrated that the bioavailability of calcium added as calcium carbonate to any of the three soy products (full-fat soy flour, soy beverage, and soy protein concentrate) was the same as when added to casein diets (Table 10). They suggested that calcium fortification of soy protein products will result in good utilization of minerals. Calcium also appears to be highly available from low phytic acid containing products such as Ca-tofu and tortillas (Table 10).

Based on results of solubility properties of calcium phytate complexes and their bioavailability to mice, Graf and Eaton[205] concluded that phytate has no detrimental effect on calcium bioavailability. Recently, Harland et al.[206] came to the same conclusions in studies, studying on the bioavailability of calcium from phytate-reduced triticale breads. They found no differences in femur calcium of rats despite significant differences in phytate intakes.

Fiber also affects the bioavailability of calcium in animals. However, the results of Weingartner et al.[84] indicate that inclusion of soybean hulls have no significant effect upon the bioavailability of the soy flour of the calcium added to rat diets (Table 10).

2. Human Studies

McCance and Widdowson[188,189] reported on the influence of dietary phytate on calcium

TABLE 10
Bioavailability of Calcium in Soy Products as Evaluated in Rats

Products	Method	Bioavailability (%)	Ref.
Full fat soy flour	Femur calcium	94	71
Soy beverage	Femur calcium	106	71
Soy protein concentrate	Femur calcium	102	71
Control, calcium carbonate	Femur calcium	100	71
Dehulled soy flour	Femur calcium	102	84
Whole soy flour	Femur calcium	108	84
Dehulled soy flour + fiber	Femur calcium	99	84
Control, calcium carbonate	Tibia calcum	100	244
Ca-tofu	Tibia calcium	107	244
Tortillas	Tibia calcium	93	244

utilization by humans and concluded that the absorption of calcium from the cereal diets could be improved by adding calcium salts or by removing phytate from the diet. Therefore, it was thought that a high concentration of phytate in cereals makes calcium unavailable for absorption by forming an insoluble calcium-phytate complex, resulting in growth reduction and even rickets. However, the effect of phytate on calcium metabolism has been regarded as temporary and the rachitogenic effect of phytate has been questioned.[190-193] It was later shown that human subjects become adapted to a high phytate diet after a short period of time and suffered no ill-effects. Whether such adaptation is due to enhanced production of intestinal phytase is not known. In several populations throughout the world normal bone and teeth calcification occurs where the diet is almost exclusively based on cereals (thus high phytate consumption), suggesting human adaptability towards high phytate consumption.[194] Several investigators[14,19,91,173,195-198] have reported the nutritional rickets and osteomalacia in the populations of northern India, Pakistan, Iran, and Bedouins and this has been attributed to a high intake of chapatis (unleavened bread) prepared from wheat flour containing high phytate and fiber. Others[199,200] however, suggest that vitamin D deficiency or the interference in vitamin D metabolism by phytate is responsible for such effects and are independent of calcium absorption in the intestine.

Anderson et al.[178] investigated the effects of breads (containing same amounts of phytate but varying levels of wheat bran) on calcium balance in humans. Three breads were prepared, equivalent to white, brown, and wholemeal, by adding wheat bran in different quantities to white flour. The phytate content of the white and brown breads was made equal to that of wholemeal bread by the addition of phytate. When the bread was fed to humans, there was no difference between the three breads for either balance or excretion of calcium. Morris et al.[201] studied the effect of three levels (0.5, 1.7, and 2.9 g/day) of phytate intake on calcium metabolic balance in humans. The mean daily intake of calcium in the diets was 728 (628 to 882) mg. Mean daily calcium balances were −23, −34, and −82 mg, respectively, for subjects consuming low, intermediate, and high phytate intake diets. They concluded that some subjects were in negative calcium balance with the low phytate intake, but high phytate intake increased the number in negative balance. In a separate study, Morris and Ellis[202] found that apparent absorption (intake-fecal excretion) of calcium decreased in individuals receiving a diet containing a molar ratio of phytate/calcium 0.14 or 0.24. The urinary excretion of calcium was not included in the calculation of apparent calcium absorption. They suggested that people consuming diets with molar ratios of phytate/calcium exceeding 0.2 may be at risk of calcium deficiency.

Van Dokkum et al.[99] observed negative mean balances for calcium in subjects consuming wheat bran breads (supplying 22 or 35 g of dietary fiber) compared to those eating white bread diet (supplying 9 g of dietary fiber). The inhibitory effects of fiber on mineral balance occur as

TABLE 11
True and Relative Availabilities of Magnesium from Foodstuffs[209]

Test ingredient	Availability True (%)	Relative index
$MgSO_4 \cdot 7H_2O$	57.4	100
Oats	82.9	144
Soybean meal (48.5% oil)	60.4	105
Wheat	56.6	99
Corn	55.9	97
Barley	54.2	95
Great Northern beans	51.1	89
Peas	48.3	84
Cream of wheat	43.8	76
Polished rice	42.2	74
Oatmeal cereal (baby food)	38.6	67
Rice cereal (baby food)	36.9	64
Minute breakfast oats	33.1	58

Note: The mean for $MgSO_4 \cdot 7H_2O$ is the average of seven experiments with four birds each. True availability is defined as the proportion (%) of dietary magnesium absorbed during the passage through the digestive tract.

a result of complexation of minerals with phytate; thus the metal salts escape absorption in the small intestine.[18,20]

D. MAGNESIUM
1. Animal Studies

Roberts and Yudkin[207] reported on dietary related magnesium deficiency symptoms in rats. The symptoms were aggravated by the addition of sodium phytate to casein-based diets. It was suggested that magnesium deficiency of unknown origin in other animals could be due to excess dietary phytate. McWard[208] reported that the addition of 4% phytic acid-soy protein complex to a purified diet containing 75 ppm of supplemental magnesium depressed chick growth and increased mortality as a result of decreased bioavailability of magnesium. Addition of 500 ppm EDTA (ethylenediaminetetraacetic acid) to a purified diet containing 4% phytic acid did not change the availability of magnesium to chicks. Nwokolo and Bragg[36] also showed that the presence of phytate decreased the availability of magnesium to growing chicks. Guenter and Sell[209] used a chick bioassay to evaluate the "true" availability of magnesium from different food stuffs as compared to $MgSO_4$. The "true" availability of magnesium from $MgSO_4$ was 57.4% and was assigned a relative availability index of 100 (Table 11). Magnesium availability in oats and soybean meal is higher than $MgSO_4$. The bioavailability of magnesium from processed cereal foods is lower than that of the whole grains from which they were prepared (Table 11).

Lo et al.[210] reported no significant difference in the bioavailability of magnesium for rats fed incremental levels of $MgCO_3$ added to isolated soy protein, casein, or lyophilized meat-based diet. In another study, Forbes et al.[71] evaluated by a slope ratio bioassay procedure the bioavailability of magnesium from three soy products (full-fat soy flour, freeze-dried soy beverage, and soy protein concentrate) as compared to $MgCO_3$ added to purified casein diets. They found that magnesium is highly available from full-fat soy flour and freeze-dried soy beverage (Table 12). Magnesium utilization from soy protein concentrate is less than from other soy products and an inorganic source, $MgCO_3$. Erdman et al.[74] observed no difference in the

TABLE 12
Bioavailability of Magnesium in Soy Products and Evaluated in Rats[71,74]

Products	Bioavailability (%)
Soy flour	106
Soy beverage (freeze-dried)	104
Soy protein concentrate	80
Soy protein concentrate (acid precipitated)	85
Soy protein concentrate (neutral)	85

Note: Availability of magnesium in soy products is relative to $MgCO_3$. Availability was measured by tibia magnesium.

bioavailability of magnesium from soy protein products prepared by two different methods (acid precipitation or acid precipitation with subsequent neutralization, Table 12). Winterringer and Ranhotra[211] reported that the supplementation of breads with soy flour or $MgCO_3$ increases relative magnesium bioavailability value for rats. Further, reducing the phytate content of cereals and cereal by-products can increase the bioavailability of magnesium in these foods.[212]

2. Human Studies

McCance and Widdowson[188] conducted magnesium balance for humans using diets containing breads (brown and white breads) prepared from different extraction wheat flours. They found low apparent absorption values for magnesium in humans who consumed a brown bread diet compared to those eating white bread diets. The poor absorption of magnesium from brown bread has been related to its phytate content. In a subsequent study, McCance and Widdowson[189] reported that absorption of magnesium from brown bread can be improved by reducing its phytate content with enzymes, i.e., dephytinization. Reinhold et al.[18] observed a negative magnesium balance in subjects consuming bazari bread which had 0.35% phytate and 3.6% fiber. The effect of three levels of phytate intake (0.5, 1.7 and 2.9 g/day) on magnesium balance in humans was investigated by Morris et al.[201] The mean daily intake of magnesium was 254 mg (ranged from 208 to 312). They[201] found that the mean balances for magnesium differed only slightly (4, 6, and –2 mg) with increasing phytate intakes. However, they observed a negative magnesium balance in some subjects receiving low phytate intake.

E. OTHER MINERALS

The effect of phytate on the absorption and availability of copper, manganese, and selenium has been studied. For example, Davis et al.[213] reported that the diets containing an isolated soy protein with phytate reduced the availability of manganese and copper for the chick. In a study with rats, Davies and Nightingale[23] found that the addition of 1% phytate to an egg albumin diet significantly reduced the absorption and availability of copper and manganese. The whole-body retention of copper was significantly reduced in rats when they received a diet containing a phytate to copper molar ratio of 40. In contrast, the results of Lee et al.[214] suggest that phytic acid increases copper bioavailability to rats by preventing zinc induction of metallothionein.

Morris and Ellis[97] found that the levels of phytate intake (0.5, 1.7, and 2.9 g daily) had no effect on the apparent absorption of copper and manganese in adult men. Turnlund et al.[215] studied the effects of phytate on copper absorption in young men. The young men were given a liquid formula diet with and without 2.35 g of phytic acid (as sodium phytate). Average copper absorption was 35.0% (with a range of 25.3 to 45.3%) from the basal diet and 31.4% (with a range of 22.0 to 44.6%) from the diet with 2.34 g of phytic acid (as sodium phytate). They concluded that the high levels of phytate do not markedly affect copper absorption. Morris et al.[216]

TABLE 13
Percent Inhibition of Alpha-Amylases Derived from Cereals and Legumes[217] by Phytic Acid at pH 4.0

Enzyme source	Phytic acid concentration (mM)			
	2	4	8	10
Wheat	25.0	52.0	78.0	84.5
Maize	24.3	53.6	83.5	91.4
Barley	22.2	55.5	67.0	—
Chick peas	21.1	55.5	77.8	—
Peanuts	17.4	34.8	45.6	54.0

Note: All measurements were made without preincubation of α-amylase with phytic acid.

determined selenium balance for young men who consumed liquid formula with and without added sodium phytate. Their[216] data indicate that sodium phytate in the liquid formula diet can increase fecal selenium excretion and thereby exert an unfavorable effect on selenium balance in humans.

III. ENZYME INHIBITION

In addition to complexing with metal ions, phytic acid also interacts with enzymes such as trypsin, pepsin, α-amylase, and β-galactosidase resulting in a decrease of their activity.[217–220,245] Sharma et al.[217] reported on phytate inhibition of α-amylases derived from cereals and legumes (Table 13). Inhibition of wheat α-amylase by phytate is reported to be noncompetitive with an apparent Ki value of 1 mM. Recently, Deshpande and Cheryan[218] found that the porcine pancreatic α-amylase activity was lowered by 16% and 95%, respectively, at 0.5 mM and 6.0 mM phytate concentrations at pH 7.0 after 15 min preincubation. The inhibition of porcine pancreatic α-amylase by phytate was stated to be a noncompetitive type with an apparent inhibitor constant of 1.75 mM.

Several mechanisms have been proposed for α-amylase inhibition by phytate. Cawley and Mitchell[219] reported that phytate suppressed α-amylase activity in sprouted wheat meal by chelating Ca^{++} which is necessary for the enzyme activity. On the contrary, Sharma et al.[217] found that addition of Ca^{++} did not reverse phytate inhibition of α-amylases. They attributed the negative effects of phytate on α-amylase activity to direct interaction of phytate with the enzyme protein at an allosteric site and not phytate complexation of Ca^{++} ions. In studies on porcine pancreatic α-amylase, Deshpande and Cheryan[218] related the inhibitory activity of phytic acid to its general complex forming ability with enzyme proteins rather than at any given site on the enzyme.

Singh and Krikorian[220] reported that *in vitro* activity of trypsin, using casein as the substrate, was substantially inhibited by low levels of phytic acid. Further *in vivo* studies need to be done on the mechanisms of inhibition of enzymes by phytic acid and/or phytates present in the foods and food products.

IV. STARCH AND PROTEIN DIGESTIBILITY

The interactions of phytic acid with α-amylase and trypsin may be important in human nutrition because such interactions may result in decreased protein and starch digestibility of foods which contain high concentrations of phytate. Phytic acid may affect starch digestibility through interaction with the amylase protein and/or binding with minerals such as calcium,

TABLE 14
Phytic Acid and Glycemic Index of Rice, Legumes, and Cereal Products[221,222,224]

Foods	Phytic acid Concentration (g/100 g, dry basis)	Intake (g/50 g)	Glycemic index
Rice	0.23	0.14	72 ± 9
Legumes			
Chick peas	0.52	0.45	36 ± 5
Black eye peas	1.38	0.83	33 ± 4
Kidney beans	1.57	1.19	29 ± 8
Kidney beans	1.15	0.87	29 ± 8
Lentils	0.73	0.75	29 ± 3
Soybeans	1.93	4.16	15 ± 5
Cereal products			
White bread	0.07	0.05	69 ± 5
Whole meal bread	0.59	0.40	72 ± 6
Cornflakes	0.09	0.05	80 ± 6
All-bran	1.54	1.69	51 ± 5
Digestive cookies	0.50	0.31	59 ± 7
Porridge (oats)	1.49	0.89	49 ± 8

whose halogen salts are known to catalyze amylase activity.[221,222] The effect of phytic acid on *in vitro* starch digestibility by human saliva enzymes and on blood glucose response (glycemic index) in healthy volunteers has been studied.[221,223] A negative significant correlation is observed between phytic acid concentration and glycemic index ($r = -0.78$, $p < 0.001$) and between phytic acid intake and glycemic index ($r = -0.71$, $p < 0.01$) in humans (Table 14). The legumes with highest phytic acid content are digested at the slowest rate and produce the lowest blood glucose responses compared to cereals and cereal products. In separate *in vitro* digestion studies involving human saliva enzymes at physiological pH and temperature, Yoon et al.[221] showed that sodium phytate (equivalent to 2% phytic acid based on the starch portion) reduced the rate of digestion of raw wheat starch by 50%. Further, they indicated that the *in vitro* phytic acid inhibition of starch digestion can be reversed by the addition of calcium. Addition of phytate to unleavened bread significantly reduces *in vitro* starch digestibility. Similarly, feeding of unleavened breads containing added phytate produces lower blood glucose responses in humans compared to unleavened breads without added phytate.[221] Knuckles and Betschart[225] evaluated the effect of phytate on *in vitro* digestion of soluble starch by human salivary α-amylase. They found that phytate at a 2 mM concentration caused a decrease of about 9% in soluble starch digestion by human salivary α-amylase. From these studies, it may be concluded that the reduced rate of starch digestibility *in vitro* and *in vivo* blood glucose response appears to be due to the presence of phytate. The effect of added phytate on the starch digestibility may be different than from the naturally occurring phytate in cereals, legumes, and their products, because cereals and legumes contain phytate in water soluble and water insoluble forms in association with protein and/or minerals. In several instances, the naturally occurring phytate forms have not been isolated and identified.[226]

Recently, researchers[227-229] studied the effect of phytate removal and readdition on the *in vitro* and *in vivo* starch digestion. They found that the removal or dephytinization of phytate produces an increase in the rate of starch digestion *in vitro* and *in vivo* blood glucose response while readdition of phytate yields an opposite effect. The effect of both endogenous and added phytic acid on *in vitro* rate of starch digestion and *in vivo* blood glucose response can be modified by the addition of calcium which binds to phytic acid and subsequently reverses the inhibitory effect.[247,248] Phytate has been suggested as a therapeutic agent to control blood glucose level in humans.[222,248] However, there are several other adverse effects of phytate that need to be considered before it can be recommended as an agent for controlling blood glucose levels in humans. Further studies should be directed towards finding appropriate levels of phytate that can lower starch digestibility and blood glucose responses in humans without causing adverse effects (mineral bioavailability).

Phytic acid also interacts with proteins to form complexes such as protein-phytate, and protein-mineral-phytate, which may result in decreased solubility, digestibility, and functionality of proteins.[226,230] *In vitro* digestibility of casein and gluten complexed with phytic acid has been shown to be lower than that of the corresponding unbound proteins.[231,232,250] Knuckles et al.[233] studied the effect of sodium phytate on pepsin digestion of casein and bovine serum albumin under *in vitro* conditions. They found that the inhibitory effect of phytate differs with substrate and increases with phytate level. In many instances, the protein source may determine the adverse effects of phytate on protein digestibility.[234,249] Several recent findings[235-239] suggest that phytate has no effect on rapeseed protein digestibility *in vitro* and *in vivo*. Data of Reddy et al.[240] also indicated that phytate has no adverse effect on *in vitro* trypsin digestibility of Great Northern bean proteins. Removal of phytate from soy protein isolates slightly improves its *in vitro* digestibility of proteins compared to control soy protein isolates.[241]

REFERENCES

1. **Oberleas, D.,** Phytates, in *Toxicants Occurring Naturally in Foods,* National Academy of Sciences, Washington, D.C., 1973, 363.
2. **O'Dell, B. L.,** Effect of soy protein on trace mineral bioavailability, in *Soy Protein and Human Nutrition,* Wilcke, H. L., Hopkins, D. T., and Waggle, D. M., Eds., Academic Press, New York, 1979, 187.
3. **Erdman, J. W., Jr.,** Oilseed phytates: nutritional implications, *J. Am. Oil Chem. Soc.,* 56, 736, 1979.
4. **Graf, E.,** Chemistry and applications of phytic acid: an overview, in *Phytic Acid: Chemistry and Applications,* Graf, E., Ed., Pilatus Press, Minneapolis, 1986, 1.
5. **Maddaiah, V. T., Kurnick, A. A., and Reid, B. L.,** Phytic acid studies, *Proc. Soc. Exp. Biol. Med.,* 115, 391, 1964.
6. **O'Dell, B. L.,** Effect of dietary components upon zinc availability: a review with original data, *Am. J. Clin. Nutr.,* 22, 1315, 1969.
7. **Vohra, P., Gray, G. A., and Kratzer, F. H.,** Phytic acid-metal complexes, *Proc. Soc. Exp. Biol. Med.,* 120, 447, 1965.
8. **Graf, E. and Eaton, J. W.,** Effects of phytate on mineral bioavailability in mice, *J. Nutr.,* 114, 1192, 1984.
9. **Wise, A.,** Dietary factors determining the biological activities of phytate, *Nutr. Abstr. Rev.,* 53, 791, 1983.
10. **Bruce, H. M. and Callow, H. K.,** Cereals and rickets: The role of inositol hexaphosphoric acid, *Biochem. J.,* 28, 57, 1934.
11. **Bronner, F., Harris, R. S., Maletskos, C. J., and Brenda, C. E.,** Studies in calcium metabolism: effect of food phytates on calcium^{45} uptake in children on low-calcium breakfasts, *J. Nutr.,* 54, 523, 1954.
12. **Krebs, H. A. and Mellanby, E.,** The effect of national wheat meal on the absorption of calcium, *Biochem. J.,* 37, 366, 1943.
13. **O'Dell, B. L., Yohe, J. M., and Savage, J. E.,** Zinc availability in the chick as affected by phytate, calcium, and ethylenediaminetetraacetate, *Poultry Sci.,* 43, 415, 1964.
14. **Reinhold, J. G., Nasr, K., Lahimgarzadeh, A., and Hedayati, H.,** Effects of purified phytate and phytate rich bread upon metabolism of zinc, calcium, phosphorus, and nitrogen in man, *Lancet,* 1, 283, 1973.
15. **Erdman, J. W., Jr., Forbes, R. M., and Kondo, H.,** Zinc bioavailability from processed soybean products, in *Nutritional Bioavailability of Zinc,* Inglett, G. E., Ed., ACS Symp. Ser. 210, American Chemical Society, Washington, D.C., 1983, 173.
16. **Morris, E. R.,** Phytate and dietary mineral bioavailability, in *Phytic Acid: Chemistry and Applications,* Graf, E., Ed., Pilatus Press, Minneapolis, 1986, 57.

17. **Rackis, J. J. and Anderson, R. L.,** Mineral availability in soy protein products, *Food Prod. Develop.*, 11, 38, 1977.
18. **Reinhold, J. G., Faradji, B., Abadi, P., and Ismail-Beigi, F.,** Decreased absorption of calcium, magnesium, zinc, and phosphorus by humans due to increased fiber, and phosphorus consumption as wheat bread, *J. Nutr.*, 106, 493, 1976.
19. **Reinhold, J. G., Parsa, A., Karimian, N., Hammick, J. W., and Ismail-Beigi, F.,** Availability of zinc in leavened and unleavened whole meal breads as measured by solubility and uptake by rat intestine *in vitro, J. Nutr.*, 104, 976, 1974.
20. **Reinhold, J. G., Ismail-Beigi, F., and Faradji, B.,** Fibre vs. phytate as determinant of the availability of calcium, zinc, and iron of bread stuffs, *Nutr. Rep. Intern.*, 12, 75, 1975.
21. **Ismail-Beigi, F., Faradji, B., and Reinhold, J. G.,** Binding of zinc and iron to wheat bread, wheat bran, and other components, *Am. J. Clin. Nutr.*, 30, 1721, 1977.
22. **Davies, N. T., Hristic, V., and Flett, A. A.,** Phytate rather than fibre in bran as the major determinant of zinc availability to rats, *Nutr. Rep. Intern.*, 15, 207, 1977.
23. **Davies, N. T. and Nightingale, R.,** The effects of phytate on intestinal absorption and secretion of zinc, and whole body retention of zinc, copper, iron, and manganese in rats, *Br. J. Nutr.*, 34, 243, 1975.
24. **Davies, N. T. and Olpin, S. E.,** Studies on the phytate: zinc molar contents in diets as a determinant of zinc availability to young rats, *Br. J. Nutr.*, 41, 590, 1979.
25. **Davies, N. T. and Reid, H.,** An evaluation of the phytate, zinc, copper, iron, and manganese contents of and zinc availability from soy-based texturized vegetable protein meat substitutes or meat extenders. *Br. J. Nutr.*, 41, 579, 1979.
26. **Tucker, H. F. and Salmon, W. D.,** Parakeratosis or zinc deficiency disease in the pig, *Proc. Soc. Exp. Biol. Med.*, 88, 613, 1955.
27. **O'Dell, B. L. and Savage, J. E.,** Symptoms of zinc deficiency in the chick, *Fed. Proc.*, 16, 394 (abstract), 1957.
28. **Morrison, A. B. and Sarett, H. P.,** Zinc deficiency in the chick, *J. Nutr.*, 65, 267, 1958.
29. **Smith, W. H., Plumlee, M. P., and Beeson, W. M.,** Effect of source of protein on zinc requirement of the growing pig, *J. Anim. Sci.*, 21, 399, 1962.
30. **Forbes, R. M.,** Excretory patterns and bone deposition of zinc, calcium, and magnesium in the rat as influenced by zinc deficiency, EDTA, and lactose, *J. Nutr.*, 74, 194, 1961.
31. **Forbes, R. M.,** Mineral utilization in the rat. III. Effects of calcium, phosphorus, lactose, and source of protein in zinc deficient and in zinc-adequate diets, *J. Nutr.*, 83, 225, 1964.
32. **Forbes, R. M. and Yohe, M.,** Zinc requirements and nitrogen balance studies with rats, *J. Nutr.*, 70, 53, 1960.
33. **Prasad, A. S., Miale, A., Jr., Farid, Z., Sandstead, H. H., Schulert, A. R., and Darby, W. J.,** Biochemical studies on dwarfism, hypogonadism, and anemia, *Arch. Int. Med.*, 111, 407, 1963.
34. **O'Dell, B. L. and Savage, J. E.,** Effect of phytic acid on zinc availability, *Proc. Soc. Exp. Biol. Med.*, 103, 304, 1960.
35. **Likuski, H. J. A. and Forbes, R. M.,** Effect of phytic acid on the availability of zinc in amino acid and casein diets fed to chicks, *J. Nutr.*, 84, 145, 1964.
36. **Nwokolo, E. N. and Bragg, D. B.,** Influence of phytic acid and crude fibre on the availability of minerals from four protein supplements in growing chicks, *Can. J. Anim. Sci.*, 57, 475, 1977.
37. **Oberleas, D., Muhrer, M. E., O'Dell, B. L., and Kintner, L. D.,** Effects of phytic acid on zinc availability in rats and swine, *J. Anim. Sci.*, 20, 945, 1961.
38. **Oberleas, D., Muhrer, M. E., and O'Dell, B. L.,** Effects of phytic acid on zinc availability and parakeratosis in swine, *J. Anim. Sci.*, 21, 57, 1962.
39. **Oberleas, D., Muhrer, M. E., and O'Dell, B. L.,** Dietary metal complexing agents and zinc availability in the rat, *J. Nutr.*, 90, 56, 1966.
40. **Likuski, H. J. A. and Forbes, R. M.,** Mineral utilization in the rat. IV. Effects of calcium and phytic acid on the utilization of dietary zinc, *J. Nutr.*, 85, 230, 1965.
41. **Makadani, D., Mickelsen, O., Ullrey, D., and Ku, P. K.,** Effect of phytic acid on rat growth and availability of zinc, copper, iron, and calcium, *Fed. Proc.*, 926, 4004 (abstract), 1975.
42. **Ranhotra, G. S., Lee, C., and Gelroth, J. A.,** Bioavailability of zinc in cookies fortified with soy and zinc, *Cereal Chem.*, 56, 552, 1979.
43. **Ranhotra, G. S., Lee, C., and Gelroth, J. A.,** Bioavailability of zinc in soy-fortified wheat bread, *Nutr. Rep. Intern.*, 18, 487, 1978.
44. **House, W. A., Welch, R. M., and Van Campen, D. R.,** Effect of phytic acid on the absorption, distribution, and endogeneous excretion of zinc in rats, *J. Nutr.*, 112, 941, 1982.
45. **Flanagan, P. R.,** A model to produce pure zinc deficiency in rats and its use to demonstrate that dietary phytate increases the excretion of endogeneous zinc, *J. Nutr.*, 114, 493, 1984.
46. **Oberleas, D.,** The effect of phytate on endogeneous zinc and zinc homeostasis, in *Trace Elements in Man and Animals — TEMA-5*, Mills, C. F., Bremner, I., and Chesters, J. K., Eds., Commonwealth Agricultural Bureaux, Aberdeen, Scotland, 1985, 453.

47. **Rader, J. I., Tao, S. H., Gaston, C. M., Wolnik, K. A., Fricke, F. L., and Fox, M. R. S.,** Effects of phytic acid on bioavailability of trace elements in soy and casein-gelatin diets fed to weanling rats, in *Trace Elements in Man and Animals — TEMA-5*, Mills, C. F., Brenmer, I., and Chesters, J. K., Eds., Commonwealth Agricultural Bureaux, Aberdeen, Scotland, 1985, 458.
48. **Yoshida, T., Shinoda, S., Kawaai, Y., Iwabuchi, A., and Mutai, M.,** The effect of gut flora on the utilization of Ca, P, and Zn in rats fed a diet containing phytate, *Agric. Biol. Chem.*, 49, 2199, 1985.
49. **O'Dell, B. L., Burpo, C. E., and Savage, J. E.,** Evaluation of zinc and availability in foodstuffs of plant and animal origin, *J. Nutr.*, 102, 653, 1972.
50. **Franz, K. B., Kennedy, B. M., and Fellers, D. A.,** Relative bioavailability of zinc from selected cereals and legumes using rat growth, *J. Nutr.*, 110, 2272, 1980.
51. **Franz, K. B., Kennedy, B. M., and Fellers, D. A.,** Relative bioavailability of zinc using weight gain of rats, *J. Nutr.*, 110, 2263, 1980.
52. **Lantzsch, H. J., Schenkel, H., and Nickerl, I.,** Zinc availability of grains and legumes, in *Trace Element Metabolism in Man and Animals — TEMA 3*, Kirchgessner, M., Ed., Freising, West Germany, 1978, 460.
53. **Welch, R. M., House, W. A., and Allaway, W. H.,** Availability of zinc from pea seeds to rats, *J. Nutr.*, 104, 733, 1974.
54. **Forbes, R. M., Erdman, J. W., Jr., Parker, H. M., Kondo, H., and Ketelsen, S. M.,** Bioavailability of zinc in coagulated soy protein (Tofu) to rats and effect of dietary calcium at a constant phytate:zinc ratio, *J. Nutr.*, 113, 205, 1983.
55. **Welch, R. M. and House, W. A.,** Availability to rats of zinc from soybean seeds as affected by maturity of seed, source of dietary protein, and soluble phytate, *J. Nutr.*, 112, 879, 1982.
56. **Forbes, R. M., Parker, H., Kondo, H., and Erdman, J. W., Jr.,** Availability to rats of zinc in green and mature soybeans, *Nutr. Res.*, 3, 699, 1983.
57. **Forbes, R. M. and Parker, H. M.,** Biological availability of zinc in and as influenced by whole fat soy flour in rat diets, *Nutr. Rep. Intern.*, 15, 681, 1977.
58. **Hettiarachchy, N. S. and Erdman, J. W., Jr.,** Bioavailability of zinc and iron from mature winged bean seed flour, *J. Food Sci.*, 49, 1132, 1984.
59. **Lantzsch, H. J. and Scheuermann, S. E.,** Effect of dehulling on zinc availability in field beans, in *Trace Element Metabolism in Man and Animals*, Gawthorne, J. M., Howell, J. M., and White, C. L., Eds., Springer-Verlag, New York, 1981, 107.
60. **Momcilovic, B., Belonji, B., Giroux, A., and Shah, B. G.,** Bioavailability of zinc in milk and soy protein based infant formulas, *J. Nutr.*, 106, 913, 1976.
61. **Beal, L., Finney, P. L., and Mehta, T.,** Effects of germination and dietary calcium on zinc bioavailability from peas, *J. Food Sci.*, 49, 637, 1984.
62. **Franz, K. B.,** Bioavailability of Zinc from Selected Cereals and Legumes, Ph.D. dissertation, University of California, Berkeley, 1978.
63. **Morris, E. R. and Ellis, R.,** Bioavailability to rats of iron and zinc in wheat bran: response to low phytate bran and effect of the phytate/zinc molar ratio, *J. Nutr.*, 110, 2000, 1980.
64. **Kratzer, F. H., Allred, J. B., Davis, P. N., Marshall, B. J., and Vohra, P.,** The effect of autoclaving soybean protein and the addition of EDTA on biological availability of dietary zinc for turkey poults, *J. Nutr.*, 68, 313, 1959.
65. **O'Dell, B. L.,** Mineral availability and metal binding constituents of the diet, *Proc. Cornell Nutr. Conference*, Ithaca, New York, 1962, 77.
66. **Lease, J. G.,** The effect of autoclaving sesame meal on its phytic acid content and on the availability of its zinc to the chick, *Poultry Sci.*, 45, 237, 1966.
67. **Lease, J. G.,** Availability to the chick of zinc-phytate complexes isolated from soy oilseed meals by an *in vitro* digestion method, *J. Nutr.*, 93, 523, 1967.
68. **Erdman, J. W., Jr.,** Bioavailability of trace minerals from cereals and legumes, *Cereal Chem.*, 58, 21, 1981.
69. **Rackis, J. J., McGhee, J. E., Honig, D. H., and Booth, A. N.,** Processing soybeans into foods: selected aspects of nutrition and flavor, *J. Am. Oil Chem. Soc.*, 52, 249A, 1975.
70. **Erdman, J. W., Jr., Weingartner, K. E., Parker, H. M., and Forbes, R. M.,** Bioavailability of calcium, zinc, and magnesium in soy protein diets, *Fed. Proc.*, 37, 891 (abstr.), 1978.
71. **Forbes, R. M., Weingartner, K. E., Parker, H. M., Bell, R. R., and Erdman, J. W., Jr.,** Bioavailability to rats of zinc, magnesium and calcium in casein-, egg-, and soy protein containing diets, *J. Nutr.*, 109, 165, 1979.
72. **Prattley, C. A., Stanley, D. W., Smith, T. K., and Van DeVoort, F. R.,** Protein-phytate interaction in soybeans. III. The effect of protein-phytate complexes on zinc bioavailability, *J. Food Biochem.*, 6, 273, 1982.
73. **Ketelsen, S. M., Stuart, M., Weaver, C. M., Forbes, R. M., and Erdman, J. W., Jr.,** Bioavailability of zinc from defatted soy flour, acid precipitated soy concentrate, and neutralized soy concentrate, as determined by intrinsic and extrinsic labeling techniques, *J. Nutr.*, 114, 536, 1984.
74. **Erdman, J. W., Jr., Weingartner, K. E., Mustakas, G. C., Schmutz, R. D., Parker, H. M., and Forbes, R. M.,** Zinc and magnesium bioavailability from acid precipitated and neutralized soybean protein products, *J. Food Sci.*, 45, 1193, 1980.

75. **Wise, A.,** Influence of calcium on trace metal-phytate interactions, in *Phytic Acid: Chemistry and Applications*, Graf, E., Ed., Pilatus Press, Minneapolis, MN, 1986, 151.
76. **Heth, D. A. and Hoekstra, W. G.,** Zinc-65 absorption and turnover in rats. I. A procedure to determine zinc-65 absorption and the antagonistic effect of calcium in a practical diet, *J. Nutr.*, 85, 367, 1965.
77. **Oberleas, D., Muhrer, M. E., and O'Dell, B. L.,** The availability of zinc from foodstuffs, in *Zinc Metabolism*, Prasad, A. S., Ed., Charles C Thomas, Springfield, Illinois, 1966, 225.
78. **Morris, E. R. and Ellis, R.,** Effect of dietary phytate/zinc molar ratio on growth and bone zinc response of rats fed semipurified diets, *J. Nutr.*, 110, 1037, 1980.
79. **Forbes, R. M., Parker, H. M., and Erdman, J. W., Jr.,** Effects of dietary phytate, calcium, and magnesium levels on zinc bioavailability to rats, *J. Nutr.*, 114, 1421, 1984.
80. **Byrd, C. A. and Matrone, G.,** Investigations of chemical basis of zinc-calcium-phytate interaction in biological systems, *Proc. Soc. Exp. Biol. Med.*, 119, 347, 1965.
81. **Oberleas, D. and Harland, B. F.,** Nutritional agents which affect metabolic zinc status, in *Zinc Metabolism: Current Aspects in Health and Disease*, Brewer, G. J., and Prasad, A. S., Eds., Alan R. Liss, New York, 1977, 11.
82. **Davies, N. T.,** Effects of phytic acid on mineral availability, in *Dietary Fiber in Health and Disease*, Vahouny, G. V. and Kritchevsky, D., Eds., Plenum Press, New York, 1982, 105.
83. **Davies, N. T.,** The effects of dietary fibre on mineral availability, *J. Plant Foods*, 3, 113, 1978.
84. **Weingartner, K. E., Erdman, J. W., Jr., Parker, H. M., and Forbes, R. M.,** Effect of soybean hull upon the bioavailability of zinc and calcium from soy flour-based diets, *Nutr. Rep. Intern.*, 19, 223, 1979.
85. **Cosgrove, D. J.,** *Inositol Phosphates: Their Chemistry, Biochemistry, and Physiology*, Elsevier, New York, 1980, 157.
86. **Reinhold, J. G.,** High phytate content of Iranian bread: A possible cause of zinc deficiency, *Am. J. Clin. Nutr.*, 24, 1204, 1971.
87. **Halsted, J. H., Ronaghy, H. A., Abadi, P., Haghshenass, M., Amirhakimi, R. M., Barakat, R. M., and Reinhold, J. G.,** Zinc deficiency in man: the Shiraz experiment, *Am. J. Med.*, 53, 277, 1972.
88. **Reinhold, J. G.,** Phytate concentrations of leavened and unleavened breads, *Ecol. Food Nutr.*, 1, 1987, 1973.
89. **Reinhold, J. G.,** Phytate destruction by yeast fermentation in whole wheat meals, *J. Am. Diet. Assoc.*, 66, 38, 1975.
90. **Reinhold, J. G.,** Zinc and mineral deficiencies in man: The phytate hypothesis, in *Review of Basic Knowledge*, Chavez, A., Bourges, H., and Basta, S., Eds., S. Karger, New York, 1975, 115.
91. **Reinhold, J. G., Faradji, B., Abadi, P., and Ismail-Beigi, F.,** An extended study of the effect of Iranian village and urban flat breads on the mineral balances of two men before and after supplementation with vitamin D, *Ecol. Food Nutr.*, 10, 169, 1981.
92. **Oberleas, D.,** Phytate content in cereals and legumes and methods of determination, *Cereal Foods World*, 28, 352, 1983.
93. **Turnlund, J. R., King, J. C., Keyes, W. R., Gong, B., and Michel, M. C.,** A stable isotope study of zinc absorption in young men: Effects of phytate and alpha-cellulose, *Am. J. Clin. Nutr.*, 40, 1071, 1984.
94. **Navert, B., Sandstrom, B., and Cederblad, A.,** Reduction of phytate content of bran by leavening in bread and its effect on zinc absorption in man, *Br. J. Nutr.*, 53, 47, 1985.
95. **Sandstrom, B., Arvidsson, B., Cederblad, A., and Rasmussen, E. B.,** Zinc absorption from composite meals. I. The significance of wheat extraction rate, zinc, calcium, and protein content in meals based on breads, *Am. J. Clin. Nutr.*, 33, 739, 1980.
96. **Sandstrom, B., Navert, B., and Cederblad, A.,** Effect of soy protein on zinc absorption from composite meals, in *Trace Elements in Man and Animals — TEMA - 5*, Mills, C. F., Bremner, S., and Chesters, J. K., Eds., Commonwealth Agricultural Bureaux, Aberdeen, Scotland, U.K., 1985, 440.
97. **Morris, E. R. and Ellis, R.,** Trace element nutriture of adult men consuming three levels of phytate, in *Trace Elements in Man and Animals — TEMA–5*, Mills, C. F., Bremner, S., and Chesters, J. K., Eds., Commonwealth Agricultural Bureaux, Aberdeen, Scotland, 1985, 443.
98. **Sandstrom, B. and Cederblad, A.,** Zinc absorption from composite meals. II. Influence of the main protein source, *Am. J. Clin. Nutr.*, 33, 1778, 1980.
99. **Van Dokkum, W., Wesstra, A., and Schippers, F. A.,** Physiological effects of fibre-rich types of bread. I. The effect of dietary fibre from bread on the mineral balance of young men, *Br. J. Nutr.*, 47, 451, 1982.
100. **Sandberg, A. S., Hasselblad, C., Hasselblad, K., and Hulten, L.,** The effect of wheat bran on the absorption of minerals in the small intestine, *Br. J. Nutr.*, 48, 185, 1982.
101. **Sandstead, H. H., Klevay, L. M., Munoz, J. M., Jacob, R. A., Logan, G. M., Jr., Dintzis, F. R., Inglett, G. E., and Shuey, W. C.,** Human zinc requirements: effects of dietary fiber on zinc metabolism, *Fed. Proc.*, 37, 254 (abstr.), 1978.
102. **Guthrie, B. E. and Robinson, M. F.,** Zinc balance studies during wheat bran supplementation, *Fed. Proc.*, 37, 254 (abstr.), 1978.

103. **Morris, E. R. and Ellis, R.,** Dietary phytate/zinc molar ratio and zinc balance in humans, in *Nutritional Bioavailability of zinc,* Inglett, G. E., Ed., ACS Symposium Series: 210, American Chemical Society, Washington, D.C., 1983, 159.
104. **Oberleas, D.,** Factors influencing availability of minerals, in *Proceedings of the Western Hemisphere Nutrition Congress IV,* White, P. L. and Selvey, N., Eds., Futura Publishing Co., Mount Kisco, 1975, 156.
105. **Lo, G. S., Settle, S. L., Steinke, F. H., and Hopkins, D. T.,** Effect of phytate: zinc molar ratio and isolated soybean protein on zinc bioavailability, *J. Nutr.,* 111, 2223, 1981.
106. **Ellis, R. and Morris, E. R.,** Relation between phytic acid and trace metals in wheat bran and soybean, *Cereal Chem.,* 58, 367, 1981.
107. **Kumar, V. and Kapoor, A. C.,** Availability of zinc as affected by phytate, *Nutr. Rep. Intern.,* 28, 103, 1983.
108. **Reddy, N. R. and Pierson, M. D.,** Isolation and partial characterization of phytic acid-rich particles from Great Northern beans *(Phaseolus vulgaris L.), J. Food Sci.,* 52, 109, 1987.
109. **Reddy, N. R. and Salunkhe, D. K.,** Interactions between phytate, protein, and minerals in whey fractions of black gram, *J. Food Sci.,* 46, 564, 1981.
110. **Reddy, N. R. and Salunkhe, D. K.,** Effects of fermentation on phytate phosphorus and minerals of black gram, rice, black gram and rice blends, *J. Food Sci.,* 45, 1708, 1980.
111. **Raboy, V., Dickinson, D. B., and Below, F. E.,** Variation in seed total phosphorus, phytic acid, zinc, calcium, magnesium, and protein among lines of *Glycine Max* and *G. Soja, Crop Sci.,* 24, 431, 1984.
112. **Beal, L. and Mehta, T.,** Zinc and phytate distribution in peas: influence of heat treatment, germination, pH, substrate, and phosphorus on pea phytate and phytase, *J. Food Sci.,* 50, 96, 1985.
113. **Kadam, S. S., Kute, L. S., Lawande, K. M., and Salunkhe, D. K.,** Changes in chemical composition of winged bean *(Psophocarpus tetragonolobus)* during seed development, *J. Food Sci.,* 47, 2051, 1982.
114. **Srikantha, S., Hettiarchchy, N. S., and Erdman, J. W., Jr.,** Nutrient, antinutrient contents, and solubility profiles of nitrogen, phytic acid and selected minerals in winged bean flour, *Cereal Chem.,* 63, 9, 1986.
115. **Tecklenburg, E., Zabik, M. E., Uebersax, M. A., Dietz, J. C., and Lucas, E. W.,** Mineral and phytic acid partitioning among air-classified bean flour fractions, *J. Food Sci.,* 49, 569, 1984.
116. **Singh, B. and Reddy, N. R.,** Phytic acid and mineral compositions of triticales, *J. Food Sci.,* 42, 1077, 1977.
117. **Oberleas, D. and Harland, B. F.,** Phytate content of foods: Effect on dietary zinc bioavailability, *J. Am. Diet. Assoc.,* 79, 433, 1981.
118. **Cerutti, G., Finoli, C., and Vecchio, A.,** Phytic acid in bran and in natural foods, *Boll. Chim. Farm.,* 123, 408, 1984.
119. **Morris, E. R. and Ellis, R.,** Phytate-zinc molar ratio of breakfast cereals and bioavailability of zinc to rats, *Cereal Chem.,* 58, 363, 1981.
120. **Harland, B. F. and Harland, J.,** Fermentative reduction of phytate in rye, white, and whole wheat breads, *Cereal Chem.,* 57, 226, 1980.
121. **Greger, J. L., Abernethy, R. P., and Bennett, O. A.,** Zinc and nitrogen balance in adolescent females fed varying levels of zinc and soy protein, *Am. J. Clin. Nutr.,* 31, 112, 1978.
122. **Young, V. R. and Janghorbani, M.,** Soy proteins in human diets in relation to bioavailability of iron and zinc: a brief overview, *Cereal Chem.,* 58, 12, 1981.
123. **Morris, E. R., Ellis, R., Hill, A. D., Steele, P., Cottrel, S., Moy, T., and Moser, P.,** Apparent zinc and iron balance of adult men consuming three levels of phytate, *Fed. Proc.,* 45, 590 (abstr.), 1986.
124. **Ellis, R., Morris, E. R., Hill, A. D., and Smith, J. C.,** Phytate: zinc molar ratio, mineral and fiber content of three hospital diets, *J. Am. Diet. Assoc.,* 81, 26, 1982.
125. **Ellis, R., Kelsay, J. L., Reynolds, R. D., Morris, E. R., Moser, P. B., and Frazier, C. W.,** Phytate:zinc and phytate calcium:zinc molar ratios in self-selected diets of Americans, Asian Indians, and Nepalese, *J. Am. Diet. Assoc.,* 87, 1043, 1987.
126. **Ellis, R., Hill, A. D., and Morris, E. R.,** Intakes of phytate, zinc, calcium, phytate:zinc and (phytate × calcium): zinc ratio by adult black Americans consuming omnivorous self-selected diets, *Fed. Proc.,* 46, 592 (abstr.), 1987.
127. **Davies, N. T., Carswell, A. J. P., and Mills, C. F.,** The effects of variation in dietary calcium intake on the phytate-zinc interactions in rats, in *Trace Elements in Man and Animals — TEMA-5,* Mills, C. F., Bremmer, I., and Chesters, J. K., Eds., Commonwealth Agricultural Bureaux, Aberdeen, Scotland, 1985, 456.
128. **Fordyce, E. J., Forbes, R. M., and Erdman, J. W., Jr.,** (Phytate) (Calcium)/(Zinc) molar ratios: are they predictive of zinc bioavailability?, *J. Food Sci.,* 52, 440, 1987.
129. **Davies, N. T. and Warrington, S.,** The phytic acid, mineral, trace element, protein and moisture content of U.K. Asian immigrant foods, *Human Nutr. Appl. Nutr.,* 40A, 49, 1986.
130. **Bindra, G. S., Gibson, R. S., and Thompson, L. U.,** (Phytate) (Calcium)/(Zinc) ratios in Asian immigrant lacto-ovo-vegetarian diets and their relationship to zinc nutriture, *Nutr. Res.,* 6, 475, 1986.
131. **Bowering, J., Sanchez, A. M., and Irwin, M. I.,** A conspectus of research on iron requirements of man, *J. Nutr.,* 106, 987, 1976.
132. **Fitch, D. C., Harville, W. E., Dinning, J. S., and Porter, F. S.,** Iron deficiency in monkeys fed diets containing soybean protein, *Proc. Soc. Exp. Biol. Med.,* 116, 130, 1964.

133. **Sathe, V. and Krishnamurthy, K.,** Phytic acid and absorption of iron, Ind. *J. Med. Res.*, 41, 453, 1953.
134. **Rotruck, J. T. and Luhrsen, K. R.,** A comparative study in rats of iron bioavailability from cooked beef and soybean protein, *J. Agric. Food Chem.*, 27, 27, 1979.
135. **Hunter, J. E.,** Iron availability and absorption in rats fed sodium phytate, *J. Nutr.*, 111, 841, 1981.
136. **Foy H., Kondi, A., and Austin, W. H.,** Effect of dietary phytate on fecal absorption of radioactive ferric chloride, *Nature*, 183, 691, 1959.
137. **Rao, K. S. and Rao, B. S. N.,** Studies on iron chelation by phytate and the influence of other mineral ions on it, *Nutr. Rep. Intern.*, 28, 771, 1983.
138. **Callender, S. T. and Warner, G. T.,** Iron absorption from brown bread, *Lancet*, 1, 546, 1970.
139. **Cowan, J. W., Esfahani, M., Salji, J. P., and Azzam, S. A.,** Effect of phytate on iron absorption in the rat, *J. Nutr.*, 90, 423, 1966.
140. **Cowan, J. W., Esfahani, M., Salji, J. P., and Nahapetian, C.,** Nutritive value of middle-eastern foodstuffs. III. Physiological availability of iron in selected foods common to middle east, *J. Sci. Food Agric.*, 18, 227, 1966.
141. **Fuhr, I. and Steenbock, H.,** The effect of dietary calcium, phosphorus, and vitamin D on the utilization of iron, *J. Biol. Chem.*, 147, 59, 1943.
142. **Ranhotra, G. S., Loewe, R. J., and Puyat, L. V.,** Effect of dietary phytic acid on the availability of iron and phosphorus, *Cereal Chem.*, 51, 323, 1974.
143. **Liebman, M. and Driskell, J.,** Dietary phytate and liver iron repletion in iron-depleted rats, *Nutr. Rep. Intern.*, 19, 281, 1979.
144. **Yoshida, T., Shinoda, S., Kawaai, Y., Iwabuchi, A., and Mutai, M.,** Gut flora and phytate for magnesium, iron, and manganese utilization in rats, *Nutr. Rep. Intern.*, 32, 1379, 1985.
145. **Patwardhan, V. N.,** The occurrence of phytin-splitting enzyme in the intestines of albino rats, *Biochem. J.*, 31, 560, 1937.
146. **Pileggi, V. J.,** Distribution of phytase in the rat, *Arch. Biochem. Biophys.*, 80, 1, 1959.
147. **Roberts, A. H. and Yudkin, J.,** Effect of phytate and other dietary factors on intestinal phytase and bone calcification in the rat, *Br. J. Nutr.*, 15, 457, 1961.
148. **Picciano, M. F., Weingartner, K. E., and Erdman, J. W., Jr.,** Relative bioavailability of dietary iron from three processed soy products, *J. Food Sci.*, 49, 1558, 1984.
149. **Morris, E. R. and Ellis, R.,** Phytate as a carrier for iron: a breakthrough in iron fortification of foods, *Bakers Digest*, 50, 28, 1976.
150. **Welch, R. M. and Campen, R. V.,** Iron availability to rats from soybeans, *J. Nutr.*, 105, 253, 1975.
151. **Monsen, E. R.,** Validation of an extrinsic iron label in monitoring absorption of nonheme food iron in normal and iron deficient rats, *J. Nutr.*, 104, 1490, 1974.
152. **Steinke, F. H. and Hopkins, D. T.,** Biological availability to the rat of intrinsic and extrinsic iron with soybean protein isolates, *J. Nutr.*, 108, 481, 1978.
153. **Schricker, B. R., Miller, D. D., and VanCampen, D.,** Effects of iron status and soy protein on iron absorption by rats, *J. Nutr.*, 113, 996, 1983.
154. **Weaver, C. M., Schmitt, H. A., Stuart, M. A., Mason, A. C., Meyer, N. R., and Elliott, J. G.,** Radiolabeled iron in soybeans: intrinsic labeling and bioavailability of iron to rats from defatted flour, *J. Nutr.*, 114, 1035, 1984.
155. **Weaver, C. M., Nelson, N., and Elliott, J. G.,** Bioavailability of iron to rats from processed soybean fractions determined by intrinsic and extrinsic labeling techniques, *J. Nutr.*, 114, 1042, 1984.
156. **Morris, E. R. and Marsh, A. C.,** Bioavailability to the growing rat of the iron of vegetables for hemoglobin formation, *Fed. Proc.*, 32, 923, (abstr.), 1973.
157. **Rodriguez, C. J., Morr, C. V., and Kunkel, M. E.,** Effect of partial phytate removal and heat upon iron bioavailability from soy protein-based diets, *J. Food Sci.*, 50, 1072, 1985.
158. **Morris, E. R. and Ellis, R.,** Isolation of monoferric phytate from wheat bran and its biological value as an iron source to the rat, *J. Nutr.*, 106, 753, 1976.
159. **May, L., Morris, E. R., and Ellis, R.,** Examination by Mossbauer spectroscopy of the chemical identity of iron phytate in wheat, *J. Agric. Food Chem.*, 28, 1004, 1980.
160. **Morris, E. R. and Ellis, R.,** Phytate, wheat bran, and bioavailability of dietary iron, in *Nutritional Bioavailability of Iron*, Kies, C., Ed., ACS Symposium Series No: 203, American Chemical Society, Washington, D.C., 1981, 121.
161. **Morris, E. R.,** An overview of current information on bioavailability of dietary iron to humans, *Fed. Proc.*, 42, 1716, 1983.
162. **Lipschitz, D. A., Simpson, K. M., Cook, J. D., and Morris, E. R.,** Absorption of monoferric phytate to dogs, *J. Nutr.*, 109, 1154, 1979.
163. **Ellis, R. and Morris, E. R.,** Effect of sodium phytate on stability of monoferric phytate complex and the bioavailability of the iron to rat, *Nutr. Rep. Intern.*, 20, 739, 1979.
164. **Widdowson, E. M. and McCance, R. A.,** Iron exchange of adults on white and brown bread diets, *Lancet*, 1, 588, 1942.

165. **McCance, R. A., Edgecombe, C. N., and Widdowson, E. M.,** Phytic acid and iron absorption, *Lancet,* 2, 126, 1943.
166. **Moore, C. V., Minnich, V., and Dubach, R.,** Absorption and therapeutic efficacy of iron phytate, *J. Am. Diet. Assoc.,* 19, 841, 1943.
167. **Sharpe, L. M., Peacock, W. C., Cooke, R., and Harris, R. S.,** The effect of phytate and other food factors on iron absorption, *J. Nutr.,* 41, 433, 1950.
168. **Hussain, R. and Patwardhan, V. N.,** Influence of phytate on absorption of iron, *Ind. J. Med. Res.,* 47, 676, 1959.
169. **Turnbull, A., Cleton, F., Finch, C. A., Thompson, L., and Martin, J.,** Iron absorption. IV. The absorption of hemoglobin iron, *J. Clin. Invest.,* 41, 1897, 1962.
170. **Apte, S. V. and Venkatachalam, P. S.,** Iron absorption in human volunteers using high phytate cereal diets, *Ind. J. Med. Res.,* 50, 516, 1962.
171. **Apte, S. V. and Venkatachalam, P. S.,** The influence of dietary calcium on absorption of iron, *Ind. J. Med Res.,* 52, 213, 1964.
172. **Elwood, P. C., Newton, D., Eakins, J. D., and Brown, D. A.,** Absorption of iron from bread, *Am. J. Clin. Nutr.,* 21, 1162, 1968.
173. **Berlyne, G. M., BenAri, J., Nord, E., and Shainkin, R.,** Bedouin osteomalacia due to calcium deprivation caused by high phytic acid content of unleavened bread, *Am. J. Clin. Nutr.,* 26, 910, 1973.
174. **Hazell, T. and Johnson, I. T.,** *In vitro* estimation of iron availability from a range of plant foods: Influence of phytate, ascorbate, and citrate, *Br. J. Nutr.,* 57, 223, 1987.
175. **Gillooly, M., Bothwell, T. H., Torrance, J. D., Macphail, A. P., Derman, D. P., Bezwoda, W. R., Mills, W., and Charlton, R. W.,** The effects of organic acids, phytates, and polyphenols on the absorption of iron from vegetables, *Br. J. Nutr.,* 49, 331, 1983.
176. **Lynch, S. R., Beard, J. L., Dassenko, S. A., and Cook, J. D.,** Iron absorption from legumes in humans, *Am. J. Clin. Nutr.,* 40, 42, 1984.
177. **Garcia-Lopez, J. S., Sherman, A. R., and Erdman, J. W., Jr.,** Do Small amounts of tannins decrease iron absorption, *Fed. Proc.,* 46, 912 (abstr.), 1987.
178. **Andersson, H., Navert, B., Bingham, S. A., Englyst, H. N., and Cummings, J. H.,** The effects of breads containing similar amounts of phytate but different amounts of wheat bran on calcium, zinc, and iron balance in man, *Br. J. Nutr.,* 50, 503, 1983.
179. **Bjorn-Rasmussen, E.,** Iron absorption from wheat bread, *Nutr. Metab.,* 16, 101, 1974.
180. **Simpson, K. M., Morris, E. R., and Cook, J. D.,** The inhibitory effect of bran on iron absorption in man, *Am. J. Clin. Nutr.,* 34, 1469, 1981.
181. **Hallberg, L., Rossander, L., and Skanberg, A. B.,** Phytates and the inhibitory effect of bran on iron absorption in man, *Am. J. Clin. Nutr.,* 45, 988, 1987.
182. **Cook, J. D., Morck, T. A., and Lynch, S. R.,** The inhibitory effect of soy products on nonheme iron absorption in man, *Am. J. Clin. Nutr.,* 34, 2622, 1981.
183. **Anon.** The effects of cereals and legumes on iron availability: a report of the international nutritional anemia consultative group, The Nutrition Foundation, Washington, D.C.
184. **Mellanby, E.,** *Spec. Rep. Ser. Med. Res. Council, London,* No. 61, 1921.
185. **Mellanby, E.,** *Spec. Rep. Ser. Med. Res. Council, London,* No. 140, 1929.
186. **Harrison, D. C. and Mellanby, E.,** Phytic acid and the rickets-producing action of cereals, *Biochem. J.,* 33, 1660, 1939.
187. **Mellanby, E.,** The rickets-producing and anticalcifying action of phytate, *J. Physiol.,* 109, 488, 1949.
188. **McCance, R. A. and Widdowson, E. M.,** Mineral metabolism of healthy adults on white and brown bread varieties, *J. Physiol.,* 101, 44, 1942.
189. **McCance, R. A. and Widdowson, E. M.,** Mineral metabolism on dephytinized bread, *J. Physiol.,* 101, 304, 1942.
190. **Cruickshank, E. W. H., Duckworth, J., Kosterlitz, H. W., and Warnock, G. W.,** The digestibility of the phytate P of oatmeal in adult man, *J. Physiol.,* 104, 41, 1945.
191. **Cullumbine, H., Basnayake, V., Lemottee, J., and Wickramanayake, T. W.,** Mineral metabolism on rice diets, *Br. J. Nutr.,* 4, 101, 1950.
192. **Walker, A. R. P., Fox, F. W., and Irving, J. T.,** Studies in human mineral metabolism: the effect of bread rich in phytate phosphorus on the metabolism of certain mineral salts with special reference to calcium, *Biochem. J.,* 42, 452, 1948.
193. **Walker, A. R. P.,** Cereals, phytic acid, and calcification, *Lancet,* 261, 244, 1951.
194. **Davidson, S. S. and Passmore, R.,** *Human Nutrition and Dietetics,* The Williams and Wilkins Co., Baltimore, Maryland, 1970, 105.
195. **Ford, J. A., Colhoun, E. M., McIntosh, W. B., and Dunnigan, M. G.,** Nutritional rickets in immigrants, *Br. Med. J.,* 2, 677, 1972.
196. **Ford, J. A., Colhoun, E. M., McIntosh, W. B., and Dunnigan, M. G.,** Biochemical response of late rickets and osteomalacia to a chapaty-free diet, *Br. Med. J.,* 3, 446, 1972.

197. **Willis, M. R., Day, R. C., Phillips, J. B., and Bateman, F. C.,** Phytic acid and nutritional rickets in immigrants, *Lancet,* 31, 771, 1972.
198. **Ford, J. A., McIntosh, W. B., and Dunnigan, M. G.,** A possible relationship between high extraction cereal and rickets and osteomalacia, *Adv. Expt. Med. Biol.,* 81, 353, 1977.
199. **Hill, L. F.,** Phytic acid and serum-calcium, *Lancet,* 2, 769, 1972.
200. **Toetia, S. P. S. and Toetia, M.,** Nutritional rickets in immigrants, *Br. Med. J.,* 2, 669, 1972.
201. **Morris, E. R., Ellis, R., Hill, A. D., Steele, P., and Cottrell, S.,** Magnesium and calcium nutriture of adult men consuming ominivore diets with three levels of phytate, *Fed. Proc.,* 44, 1675 (abstr.), 1985.
202. **Morris, E. R. and Ellis, R.,** Bioavailability of dietary calcium: Effect of phytate on adult men consuming nonvegetarian diets, in *Nutritional Bioavailability of Calcium,* Kies, C., Ed., ACS Symposium Series 275, American Chemical Society, Washington, D.C., 1985, 63.
203. **Van Den Berg, C. J., Hill, C. F., and Stanbury, S. W.,** Inositol phosphates and phytic acid as inhibitors of biological calcification in the rat, *Clin. Sci.,* 43, 377, 1972.
204. **Thomas, W. C., Jr. and Tilden, M. T.,** Inhibition of mineralization by hydrolyzates of phytic acid, *Johns Hopkins Med. J.,* 131, 133, 1972.
205. **Graf, E. and Eaton, J. W.,** Dietary phytate and calcium bioavailability, in *Nutritional Bioavailability of Calcium,* Kies, C., Ed., ACS Symposium Series 275, American Chemical Society, Washington, D.C., 1985, 51.
206. **Harland, B. F., Felix-Phipps, R., Knight, E. M., Oke, O. L., and Adkins, J. S.,** Effect of calcium and zinc status in rats fed phytate-reduced triticale breads, *Fed. Proc.,* 46, 600 (abstr.), 1987.
207. **Roberts, A. H. and Yudkin, J.,** Dietary phytate as a possible cause of magnesium deficiency, *Nature,* 185, 823, 1960.
208. **McWard, G. W.,** Effects of phytic acid and ethylenediaminetetraacetic acid (EDTA) on the chick's requirement for magnesium, *Poultry Sci.,* 48, 791, 1969.
209. **Guenter, W. and Sell, J. L.,** A method for determining "true" availability of magnesium from foodstuffs using chickens, *J. Nutr.,* 104, 1446, 1974.
210. **Lo, G. S., Collins, D. W., Steinke, F. H., and Hopkins, D. T.,** Effect of isolated soy protein on bioavailability of magnesium, *Fed. Proc.,* 37, 667 (abstr.), 1978.
211. **Winterringer, G. L. and Ranhotra, G. S.,** Relative bioavailability of magnesium from mineral- and soy-fortified breads, *Cereal Chem.,* 60, 14, 1983.
212. **Harmuth-Hoene, V. A. E.,** The bioavailability of magnesium in wheat bran with differing phytate content, *Magnesium Bull.,* 7, 29, 1985.
213. **Davis, P. N., Norris, L. C., and Kratzer, F. H.,** Interference of soybean proteins with the utilization of trace minerals, *J. Nutr.,* 77, 217, 1961.
214. **Lee, D. Y., Schroeder, J. J., Hess, R. L., and Gordon, D. T.,** Interaction of phytic acid and zinc on intestinal metallothionein levels and copper bioavailability, *Fed. Proc.,* 44, 1674 (abstr.), 1985.
215. **Turnlund, J. R., King, J. C., Gong, B., Keyes, W. R., and Michel, M. C.,** A stable isotope study of copper absorption in young men: effect of phytate and alpha-cellulose, *Am. J. Clin. Nutr.,* 42, 18, 1985.
216. **Morris, V. C., Turnlund, J. R., King, J. C., and Levander, O. A.,** Effect of sodium phytate and alpha-cellulose on selenium balance in young men, *Fed. Proc.,* 44, 1670 (abstr.), 1985.
217. **Sharma, C. B., Goel, M., and Irshad, M.,** Myoinositol hexaphosphate as a potential inhibitor of alpha-amylases, *Phytochemistry,* 17, 201, 1978.
218. **Deshpande, S. S. and Cheryan, M.,** Effects of phytic acid, divalent cations, and their interactions on alpha-amylase activity, *J. Food Sci.,* 49, 516, 1984.
219. **Cawley, R. W. and Mitchell, T. A.,** Inhibition of wheat alpha-amylase by bran phytic acid, *J. Sci. Food Agric.,* 19, 106, 1968.
220. **Singh, M. and Krikorian, A. D.,** Inhibition of trypsin activity *in vitro* by phytate, *J. Agric. Food Chem.,* 30, 799, 1982.
221. **Yoon, J. H., Thompson, L. U., and Jenkins, D. J. A.,** The effect of phytic acid on *in vitro* rate of starch digestibility and blood glucose response, *Am. J. Clin. Nutr.,* 38, 835, 1983.
222. **Thompson, L. U.,** Phytic acid: a factor influencing starch digestibility and blood glucose response in *Phytic Acid: Chemistry and Applications,* Graf, E., Ed., Pilatus Press, Minneapolis, 1986, 173.
223. **Thompson, L. U. and Yoon, J. H.,** Starch digestibility as affected by polyphenols and phytic acid, *J. Food Sci.,* 49, 1228, 1984.
224. **Jenkins, D. J. A., Wolever, T. M. S., Taylor, R. H., Barker, H. M., Fielden, H., Baldwin, J. M., Bowling, A. C., Newman, H. C., Jenkins, A. L., and Goff, D. V.,** Glycemic index of foods: a physiological basis for carbohydrate exchange, *Am. J. Clin Nutr.,* 34, 362, 1981.
225. **Knuckles, B. E. and Betschart, A. A.,** Effect of phytate and other inositol phosphate esters on alpha-amylase digestion of starch (abstr.), *46th IFT Ann. Meet.,* Dallas, 1986.
226. **Reddy, N. R., Sathe, S. K., and Salunkhe, D. K.,** Phytates in legumes and cereals, *Adv. Food Res.,* 28, 1, 1982.
227. **Button, C. L., Thompson, L. U., and Jenkins, D. J. A.,** Activities of indigenous phytic acid and added sodium phytate in lowering the digestibility of legumes, *Proc. Can. Fed. Biol. Soc.,* 27, 73 (abstr.), 1984.

228. **Button, C. L., Thompson, L. U., and Jenkins, D. J. A.,** Influence of phytic acid and calcium on starch digestibility, glycemic response, and carbohydrate malabsorption, *Proc. Can. Fed. Biol. Soc.,* 28, 137 (abstr.), 1985.
229. **Thompson, L. U., Button, C. L., and Jenkins, D. J. A.,** Phytic acid as a determinant of starch digestibility and blood glucose response, *Proc. Intern. Congr. Nutr.* (abstr.), 1985, 2.
230. **Cheryan, M.,** Phytic acid interactions in food systems, *CRC Crit. Rev. Food Sci. Nutr.,* 13, 297, 1980.
231. **Barre, R. and Nguyen Vanm Huot, N.,** Etude de la combinaison de l'ovalbumine avec les acids phosphorique, P-glycerophos phorique et phytique, *Bull. Soc. Chim. Biol.,* 47, 1419, 1965.
232. **Camus, M. C. and Laporte, J. C.,** Inhibition de la proteolyse pepsique *in vitro* par le blé: Role de l'acide phytique des issues, *Ann. Biol. Anim. Biochim. Biophys.,* 16, 719, 1976.
233. **Knuckles, B. E., Kuzmicky, D. D., and Betschart, A. A.,** Effect of phytate and partially hydrolyzed phytate on *in vitro* protein digestibility, *J. Food Sci.,* 50, 1080, 1985.
234. **Satterlee, L. D. and Abdul-Kadir, R.,** Effect of phytate content on protein nutritional quality of soy and wheat bran, *Lebenson. Wiss. Technol.,* 16, 8, 1983.
235. **Lieden, S. A. and Hambraeus, L.,** Removal from rapeseed of a low molecular weight substance affecting the pregnant rat, *Nutr. Rep. Intern.,* 10, 367, 1977.
236. **McDonald, B. E., Lieden, S. A., and Hambraeus, L.,** Evaluation of the protein quality of rapeseed meals, flours, and isolates, *Nutr. Rep. Intern.,* 17, 49, 1978.
237. **McDonald, B. E., Lieden, S. A., and Anjou, K.,** Nutritional assessment of rapeseed protein concentrate as a meat extender, *Nutr. Rep. Intern.,* 18, 51, 1978.
238. **Serraino, M. R., Thompson, L. U., Savoie, L., and Parent, G.,** Effect of phytate on the *in vitro* digestibility of rapeseed protein and amino acid, *J. Food Sci.,* 50, 1689, 1985.
239. **Thompson, L. U. and Serriano, M. R.,** Effect of phytate reduction on rapeseed protein digestibility and amino acid absorption, *J. Agric. Food Chem.,* 34, 468, 1986.
240. **Reddy, N. R., Sathe, S. K., and Pierson, M. D.,** Removal of phytate from Great Northern beans *(Phaseolus vulgaris L.)* and its combined density fraction, *J. Food Sci.,* 53, 107, 1988.
241. **Ritter, M. A., Morr, C. V., and Thomas, R. L.,** *In vitro* digestibility of phytate-reduced and phenolics-reduced soy protein isolates, *J. Food Sci.,* 52, 325, 1987.
242. **Thompson, D. B. and Fosmire, G.J.,** Determination of zinc bioavailability from soy protein concentrates by slope-ratio analysis, *J. Food Sci.,* 53, 204, 1988.
243. **Beard, J. L., Weaver, C. M., Lynch, S., Johnson, C. D., Dassenko, S., and Cook, J. D.,** The effect of soybean phosphate and phytate content on iron bioavailability, *Nutr. Res.,* 8, 345, 1988.
244. **Poneros, A. G. and Erdman, J. W., Jr.,** Bioavailability of calcium from tofu, tortillas, nonfat dry milk, and mozzarella cheese in rats: effect of supplemental ascorbic acid, *J. Food Sci.,* 53, 208, 1988.
245. **Inagawa, J., Kiyosawa, I., and Nagasawa, T.,** Effect of phytic acid on the hydrolysis of lactose with beta-galactosidase, *Agric. Biol. Chem.,* 51, 3027, 1987.
246. **Harland, B. F. and Oberleas, D.,** Phytate in foods, *World Rev. Nutr. Diet.,* 52, 235, 1987.
247. **Thompson, L. U., Button, C. L., and Jenkins, D. J. A.,** Phytic acid and calcium affect the *in vitro* rate of navy bean starch digestion and blood glucose response in humans, *Am. J. Clin. Nutr.,* 46, 467, 1987.
248. **Thompson, L. U.,** Antinutrients and blood glucose, *Food Technol.,* 42, 123, 1988.
249. **Carnovale, E., Lugaro, E., and Lombardi-Boccia, G.,** Phytic acid in faba beans and pea: effect on protein availability, *Cereal Chem.,* 65, 114, 1988.
250. **Lathia, D., Hoch, G., and Kievernagel, Y.,** Influence of phytate on *in vitro* digestibility of casein under physiological conditions, *Qual. Plant. Plant Foods Hum. Nutr.,* 37, 229, 1987.

Chapter 10

INFLUENCE OF PROCESSING TECHNOLOGIES ON PHYTATE

I. INTRODUCTION

Cereals and legumes are processed differently and consumed in a variety of forms throughout the world. These foods are an important source of calories and protein (quality and quantity) to many populations of the world. Cereals are either wet- or dry-milled to produce a diverse group of edible byproducts that can be used in the preparation of human food products and animal feed components. Flour is used by the baking industry to make bread, cookies, cakes, etc. Whole cereals and part of the byproducts are used in preparation of breakfast foods. The most common modes of preparation of legumes include cooking, germination (sprouting), fermentation, and roasting/frying.[1] Cooking is one of the most prevalent methods used for preparation of legumes for consumption.

Soaking the legume overnight (10 to 12 h) in water is the first step before cooking of legumes. The cooking time ranges up to 4 h, depending upon the type of legume under consideration. Attempts to eliminate or reduce phytate in legumes, cereals, and their products by processing methods (cooking, soaking, germination, fermentation, etc.), have been partially successful. Reduction of phytate in legumes and cereals during food processing may increase the digestibility of proteins and bioavailability of essential minerals.

II. COOKING

The cookability can be defined as the cooking time required for beans to reach the cooked texture that is considered acceptable for eating.[2] Generally, the time required to soften freshly harvested beans is taken as a basis for comparison. A correlation between cookability and phytate content of peas was first suggested by Mattson.[3] He showed that the cooking quality of peas deteriorates when the phytate level in peas is low. Further, Mattson[3] hypothesized that phytate chelated calcium and magnesium ions, preventing formation of calcium and magnesium cross-linkages between the pectate molecules of the middle lamellar tissue of peas. Higher phytate content in beans favors a rapid rate of softening and dissolution of pectic substances that would result in shorter cooking times. On the other hand, lower phytate content in beans permits more interpectate cross-link formation that would result in a slower dissolution of pecticsubstances and in longer cooking times.[2] Moscoso et al.[2] suggested that phytate content can be used as an index of cookability in beans. Several earlier and recent studies, covering many species and varieties of legumes, support the existence of a relationship between phytate content and cookability.[2,4-13,87] Other researchers[14-17] failed to obtain a good correlation between these two parameters.

Crean and Haisman[14] investigated the interactions between phytate and divalent cations (Ca and Mg) and the rate of conversion of soluble phytate to its insoluble form in dry peas during cooking in water with and without the addition of calcium chloride. They found that phytate in dry peas exists entirely as a water-soluble salt (probably as potassium phytate) but, on cooking, some of it combines with the calcium and magnesium in the pea to form insoluble calcium and magnesium phytate. At high concentrations of calcium chloride only 60% of the available phytate in the peas is converted to insoluble form (Figure 1). Crean and Haisman[14] obtained similar results with magnesium chloride, a maximum of 43% of the phytate being complexed when the peas were cooked in high concentrations of magnesium chloride. Based on these observations, Crean and Haisman[14] concluded that the influence of phytate on the texture of peas is small and a high phytate content is an unreliable criterion of good cooking quality. Rosenbaum

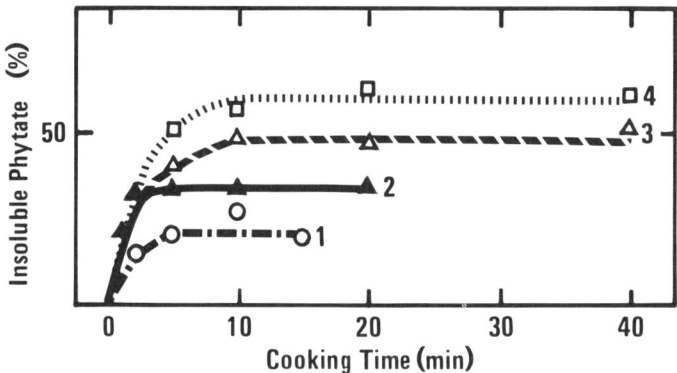

FIGURE 1. Rate of formation of insoluble phytate in peas cooked in calcium chloride solutions. Calcium ions added to the cooking liquid equivalent $\times 10^4$ per 10 g dry peas. (1) 0 (2) 2.44 (3) 4.88, and (4) 9.76. (From Crean, D. E. C. and Haisman, D. R., *J. Sci. Food Agric.*, 14, 824, 1963. With permission.)

TABLE 1
Effects of Cooking on Water and Acid Extractable Phytate in Green Gram, Cowpea, and Chick Peas[7]

Legume	Phytate (%)			
	Water extracted	Retained	Acid extracted	Retained
Green gram				
Uncooked	0.50	100.0	0.67	100.0
Cooked	0.28	56.0	0.53	79.1
Cowpea				
Uncooked	0.32	100.0	0.44	100.0
Cooked	0.11	34.3	0.32	72.7
Chick pea				
Uncooked	0.20	100.0	0.28	100.0
Cooked	0.13	65.0	0.27	96.4

Note: Original phytate phosphorus values were converted to phytate, assuming that phytate contains 28.20% phosphorus.

and Baker[15] also reported that the higher cookability of the pea is not associated with a higher phytate content. Cooking time of beans containing low amounts of phytate can be reduced by soaking them in a phytate solution.[9]

In green gram, cowpea, and chick pea, the cooking process decreases both the water and acid-extractable phytate content (Table 1).[7] A small change in the ratio of phytate phosphorus/total phosphorus in the uncooked and cooked legumes (in only acid-extracted legumes) was observed by Kumar et al.[7] The poor water and acid extractability of phytate in all three cooked legumes could be due to formation during cooking of insoluble complexes between phytate and other components in legumes which subsequently cannot be extracted with water or acid.

Reddy et al.[18] did not find a breakdown of phytate during cooking of black gram seeds and cotyledons (Figure 2). Whatever losses in total phosphorus and phytate they observed during short time cooking were due to leaching of those components into the cooked water. Cooking for 45 min at 115°C caused small losses in total phosphorus and phytate into cooked water, which may have been due to reabsorption of phytate by beans from cooked water. In black gram, more

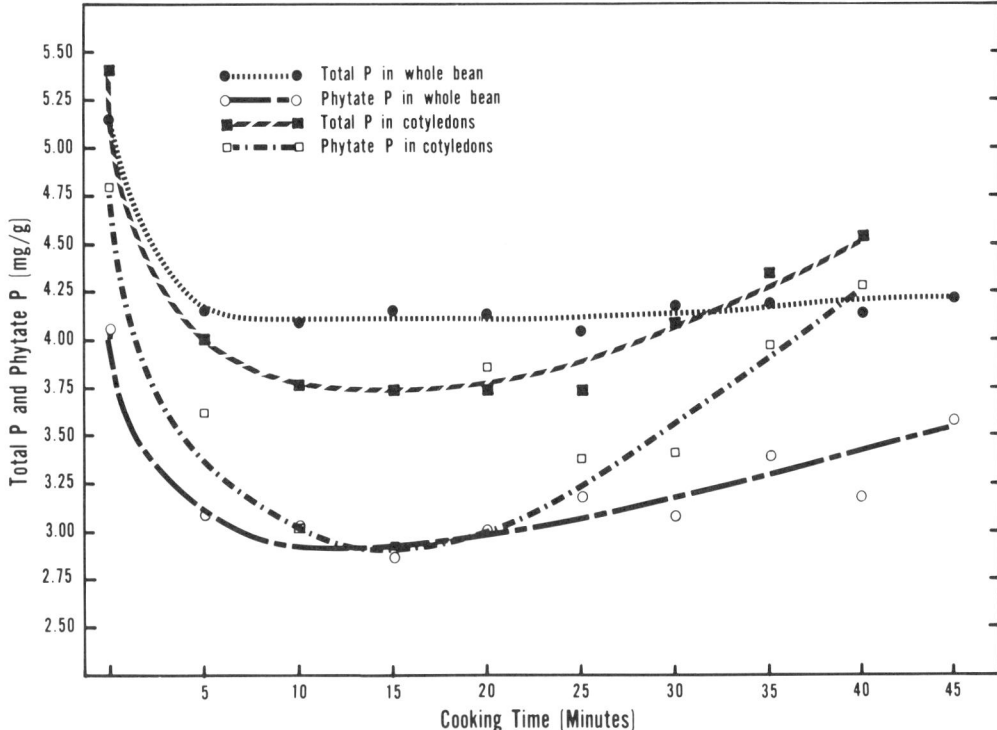

FIGURE 2. Effects of cooking on total P and phytate P contents in black gram seeds and cotyledons. Cooking was carried out at 10 psi (116°C) for different time lengths (bean to water ratio of 1:4). At the end of cooking, cookwater was discarded and samples were immediately lyophylized for phytate P and total P analyses. (From Reddy, N. R., Balakrishnan, C. V., and Salunkhe, D. K., *J. Food Sci.*, 43, 540, 1978. With permission.)

than 80% of phytate exists in water-soluble form.[19] This further supports the findings of Reddy et al.[18] About 58 to 85% of the phytate remains in mung beans after autoclaving at 120°C for 30 minutes.[20] The researchers in this report did not indicate whether the cooked water is discarded or included in the analysis. Tabekhia and Luh[21] found a slight decrease in phytate content of black-eyed and red kidney beans cooked in water containing 3% salt (beans:water 1:4 w/v) compared to mung and pink beans (Table 2). Canning beans with 3% salt water at 115.5° C for 3 h results in a significant reduction of phytate in all four beans. The retention of phytate in the canned black eyed beans is only 8.50%, whereas that in the other three beans (red kidney, mung, and pink beans) ranges from 25.1 to 32.4% (Table 2). Iyer et al.[22] observed a reduction in the phytate content of pinto, Great Northern, and red kidney beans during a combined process of soaking and cooking (Table 3). Hydrolysis of phytate is higher in beans soaked and cooked in distilled water than those soaked in salt solution and cooked. The salt solution-soaked and cooked beans retain 70 to 80% of the original phytate. Several others[12,23] have reported lower phytate reductions in beans soaked in salt solutions and cooked. The cooking process also reduces phytate by 13% in peas[24] and less than 15% in cowpeas and lima beans.[25] Clydesdale and Camire[26] studied the effect of heat treatment on phytate destruction in soy flour. They obtained a greater reduction of phytate in boiled soy flours at pH 5.0 and 6.8 compared to toasted soy flour (Table 4). Boiling results in a 22% and 32% loss at pH 6.8 and 5.0, respectively, in soy flour.

Lease[27] found no decrease in the phytate content of sesame meal after autoclaving for 2 h and only a 22% decrease after 4 h (Table 5). de Boland et al.[28] studied the rate of destruction of phytate during heat processing (autoclaving) of cereal and oil seed products and inositol hexaphosphate. They reported that the amount of water added during autoclaving has little or no effect on the

TABLE 2
Effects of Cooking and Canning on Phytate Retention in Beans[21]

Bean type	Phytate content (%)	Phytate retained (%)
Black eyed beans		
Raw dry beans	1.15	100.0
Cooked beans	1.00	87.0
Canned beans	0.10	8.7
Red kidney beans		
Raw dry beans	1.17	100.0
Cooked beans	1.08	92.3
Canned beans	0.35	30.3
Mung beans		
Raw dry beans	0.204	100.0
Cooked beans	0.130	63.7
Canned beans	0.066	32.4
Pink beans		
Raw dry beans	0.503	100.0
Cooked beans	0.370	73.6
Canned beans	0.126	25.1

TABLE 3
Effects of Soaking and Cooking on Phytate Concentrations in Great Northern, Pinto, and Red Kidney Beans[23]

	Great Northern		Pinto		Red kidney	
Treatment	Phytate content (mg/g)	Phytate hydrolyzed (%)	Phytate content (mg/g)	Phytate hydrolyzed (%)	Phytate content (mg/g)	Phytate hydrolyzed (%)
Control (Raw dry beans)	16.3	—	19.5	—	20.6	—
Soaking[a] + cooking	3.9	76.1	7.5	61.5	8.9	57.8
Soaking[b] + cooking	14.2	13.0	14.5	25.6	14.2	31.3

[a] Soaked for 18 h in distilled water at room temperature before quick cooking.
[b] Soaked for 18 h in a mixed salt solution (2.5% sodium chloride, 1.5% sodium bicarbonate, 0.5% sodium carbonate, and 1.0% sodium tripolyphosphate in distilled water), pH 7.0 at room temperature before quick cooking.

rate of phytate destruction. Results show that 30 min autoclaving reduces the phytate content of cereals and oil seed products by less than 10% (Figure 3). Autoclaving inositol hexaphosphate in aqueous solution (pH 6.0) results in nearly an 80% loss of iron-precipitable phosphorus in 2 h, and approximately 50% is lost within 1 h. The next most labile source of phytate is that in isolated soy protein — nearly 70% loss in 2 h (Figure 3). Phytate in other products is relatively stable. Losses of phytate varies from 25% in sesame meal to 5% in rice and wheat in 2 h at 115°C. These results suggest that the rate of phytate destruction, probably by hydrolysis, is influenced by its protein and/or cation environment. Chompreeda and Fields[90] reported that the autoclaving (121°C for 30 min) of soybean meal results in a 17.5% reduction of phytate.

Camire and Clydesdale[29] found that heating of wheat bran for 1 h in a boiling water bath had no effect on phytate destruction, whereas toasting of wheat bran results in a 19% loss of phytate.

TABLE 4
Effects of pH and Wet and Dry Heat Treatment on the Retention of Phytate in Soy Flour[26]

Sample	Treatment	Phytate (%) Retention	Loss
Defatted soy flour[a]	Control	100	0
	Toast (1 h at 177°C)	88	12
	Boil (1 h, BWB, pH 6.80)	78	22
	Boil (1 h, BWB, pH 5.0)	68	32

Note: BWB =Boiling water bath. Experimental conditions: toasting: samples weighed into aluminum foil boats and placed in a preheated oven at 177°C for 1 h. Boiling: samples (1 g) weighed into screw cap centrifuge tubes, to which was added about 30 ml of either pH 5.0 buffer (phthalate-sodium hydroxide) or double distilled deionized water at an unbuffered pH of 6.80. The samples were mixed, capped and placed in a boiling water bath for 1 h.

[a] Soy flour contained 22.5 mg/g of phytate

TABLE 5
Effects of Autoclaving on the Phytic Acid Content of Sesame Meal[27]

Hours at 15 psi	Phytic acid phosphorus (%)	Phytic acid Content (%)	Retained (%)
0	1.00	3.55	100.0
2	1.03	3.65	103.0
4	0.78	2.77	77.9

Tangendjaja et al.[30] obtained a maximum reduction (more than 75%) in phytate level of rice bran by incubating the bran at 55°C and pH 5.10 for 24 h. Both steaming rice bran for 1 h and autoclaving rice bran for 1 h at 121°C reduced phytate level by 25%. The studies of Toma and Tabekhia[31] showed that cooking milled rice in domestic tap water reduces phytate content by about 70% (Table 6), whereas cooking milled rice in distilled deionized water did not reduce phytate content. Fretzdorff and Weipert[89] reported that extrusion cooking of whole rye flour at up to 100°C did not influence the phytate level; however, there was a 23% reduction at 170°C.

It appears that destruction and/or reduction of phytate in cereals, legumes, and their products during cooking is dependent on factors such as cooking conditions, temperature and pH, product type, presence of phytate form in the product under consideration, and presence of proteins and/or associated cations. For instance, the rate of destruction of phytate by autoclaving is low when phytate is associated with proteins and/or cations in natural products such as cereal and/or cations in natural products such as cereal and oil seed products.

III. GERMINATION

Phytate is utilized as a source of inorganic phosphate during seed germination and inorganic phosphate becomes available for purposes of plant growth and development. During germination, the hydrolysis of phytate in the seed is catalyzed by the enzyme phytase to inositol and free orthophosphate. Presence of phytase in cereals and legumes has been reported and its activity

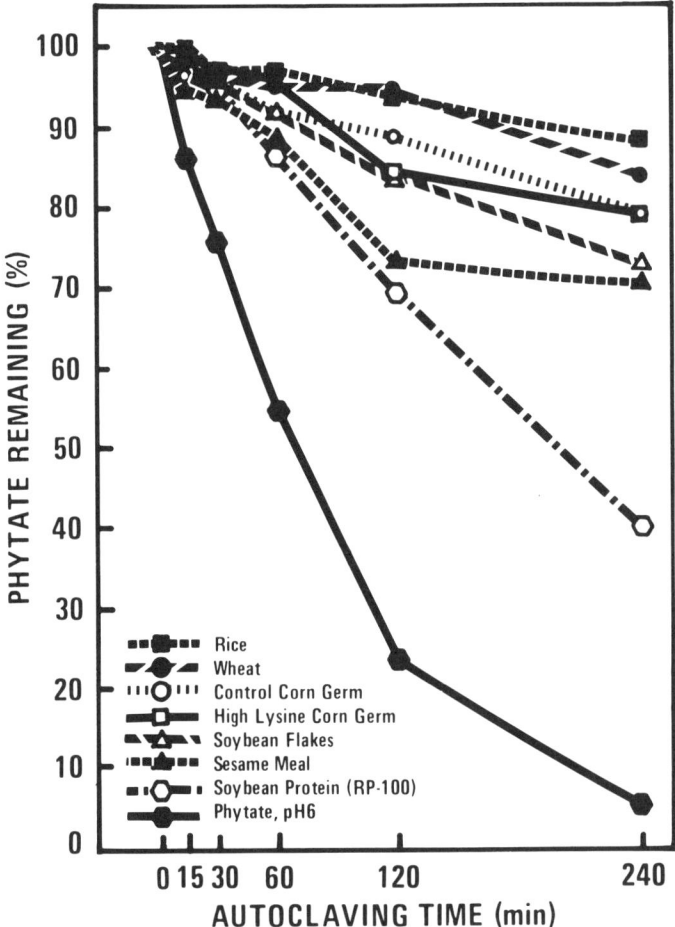

FIGURE 3. Rate of phytate loss during autoclaving phytate and moist slurries of various cereal and oil seed products at 115°C. (From deBoland, A., Garner, G. B., and O'Dell, B. L., *J. Agric. Food Chem.*, 23, 1186, 1975. With permission from the American Chemical Society.)

varies widely. Oats, maize, and some millets have negligible phytase activity; barley has moderate activity; wheat and rye have high activity.[32] Belavady and Banerjee[33] reported an absence of phytase activity in some ungerminated legumes. However, they found phytase activity in germinated legumes. Phytase activity increases during germination of cereals and legumes.[7,24,34-43,88]

Germination can lower the phytate content in legumes and cereals to a variable degree depending on the germinating conditions and type of legume or cereal. Germination also improves the availability of certain minerals through phytate hydrolysis by phytase. The percentages of phytate hydrolyzed at various stages of germination of different beans are presented in Table 7. During germination, hydrolysis of phytate in beans varies greatly. The percentage of phytate hydrolyzed in beans ranges from 20.2 to 77.4% after germination for 5 days. In some beans, very little phytate hydrolysis occurs during the first 2 days of germination.[23,45] The maximum amount (77.4%) of phytate hydrolysis is in black eyed beans during 5-day germination (Table 7). In most beans, complete hydrolysis of phytate does not take place during a 5-day germination. More than 70% of the phytate is hydrolyzed in peas and faba beans that are germinated for 10 days.[24,43] Disappearance of phytate during germination of beans depends on the phytase activity. Walker[38] found no change in phytate content and phytase

TABLE 6
Effects of Cooking on Phytic Acid Retention in Milled Rice Varieties[31]

Rice variety	Phytic acid Content (%)	Retained (%)
Terso		
Raw rice	0.192	100.0
Rice cooked with DDW[a]	0.188	97.9
Rice cooked with DTW[b]	0.054	28.1
M-5		
Raw rice	0.140	100.0
Rice cooked with DDW	0.135	96.4
Rice cooked with DTW	0.045	32.1
S-6		
Raw rice	0.137	100.0
Rice cooked with DDW	0.135	98.5
Rice cooked with DTW	0.042	30.7

[a] DDW = distilled deionized water.
[b] DTW = domestic tap water.

TABLE 7
Effect of Germination on Phytate Content of Legumes

Legume	Days	Phytate phosphorus (mg/g)	Phytate (mg/g)	% Phytate hydrolyzed	Ref.
Pigeon pea	0	0.35	1.24	—	3
	1	0.34	1.21	2.4	
	3	0.34	1.21	2.4	
	5	0.28	0.99	20.2	
Chick pea	0	1.24	4.40	—	33
	1	1.21	4.30	2.3	
	3	1.11	3.94	10.5	
	5	0.73	2.59	41.1	
Lentils	0	0.87	3.09	—	33
	1	0.88	3.12	—	
	3	0.67	2.38	23.0	
	5	0.41	1.46	52.8	
Cowpea	0	0.62	2.20	—	33
	1	0.59	2.09	5.0	
	3	0.41	1.46	33.6	
	5	0.31	1.10	50.0	
Cowpea	0	4.08	14.50	—	25
	3	2.43	8.60	40.7	
Garden pea	0	0.91	3.23	—	33
	1	0.62	2.20	32.0	
	3	0.42	1.49	53.9	
	5	0.28	0.99	69.4	
Dwarf grey pea	0	2.48	8.80	—	39
	5	1.94	6.89	21.7	
Early Alaska pea	0	1.13	4.01	—	39
	5	0.59	2.09	47.9	

TABLE 7 (continued)
Effect of Germination on Phytate Content of Legumes

Legume	Days	Phytate phosphorus (mg/g)	Phytate (mg/g)	% Phytate hydrolyzed	Ref.
Soybean	0	1.86	6.60	—	39
	5	1.20	4.26	35.6	
Black gram	0	0.81	2.88	—	33
	1	0.74	2.63	8.7	
	3	0.45	1.60	44.4	
	5	0.32	1.14	60.4	
Black gram	0	4.10	14.56	—	18
	1	3.70	13.14	9.8	
	3	3.40	12.07	17.1	
	5	3.10	11.01	24.4	
	7	2.80	9.94	31.7	
	10	2.00	7.10	51.2	
Green gram	0	0.88	3.12	—	33
	1	0.81	2.88	7.7	
	3	0.38	1.35	5.7	
	5	0.36	1.28	59.0	
Green gram	0	0.58	2.05	—	21
	1	0.48	1.70	17.1	
	3	0.40	1.42	30.7	
	5	0.40	1.43	30.2	
Black eyed beans	0	3.23	11.48	—	21
	1	2.99	10.92	4.9	
	3	2.03	7.20	37.3	
	5	0.73	2.59	77.4	
Lima bean	0	3.02	11.07	—	25
	3	2.03	7.20	32.7	
Red kidney bean	0	3.30	11.70	—	21
	1	2.99	10.63	9.2	
	3	2.18	7.73	33.9	
	5	2.11	7.50	35.9	
Pink bean	0	1.42	5.03	—	21
	1	1.23	4.38	12.9	
	3	0.96	3.40	32.4	
	5	0.76	2.69	46.5	
Great Northern bean	0	5.10	18.10	—	44
	2	4.03	14.30	21.0	
	3	3.13	11.10	38.7	
	5	2.14	7.60	57.8	

activity of kidney beans during the initial 2 days of germination; however, a 50% phytate loss followed by day 5. Chen and Pan[39] reported an increase of 227, 807, and 3756% of phytase activity, respectively, in soy beans, dwarf grey peas, and early Alaska peas after 5-day germination. A correlation exists between phytate breakdown and phytase activity of the seeds during germination. Such correlations have been demonstrated for kidney beans,[38] mung beans,[46] navy beans,[40] early Alaska pea, dwarf grey pea, and soy beans.[39]

The effects of germination on the extent of phytate hydrolysis in various cereal grains are presented in Table 8. All the grains were steeped for 2 days at room temperature and then germinated at 25°C. Except in the case of rye, complete hydrolysis of phytate did not occur during germination for up to 7 days. On the seventh day of germination, wheat, yellow corn, white corn, barley, and oats still retained 47.4, 33.0, 50.0, 34.8, and 67.2% of the original phytate respectively (Table 8). Fretzdorff and Weipert[89] found no change in phytase activity during early

TABLE 8
Effect of Germination on Phytate Content of Cereals

Cereal	Days	Phytate phosphorus (mg/g)	Phytate (mg/g)	% Phytate hydrolyzed	Ref.
Wheat	0	2.32	8.24	—	47
	1	2.00	7.10	13.8	
	3	1.61	5.72	30.6	
	5	1.39	4.93	40.1	
	7	1.10	3.91	52.6	
Yellow corn	0	2.12	7.53	—	47
	1	1.64	5.82	22.6	
	3	1.17	4.15	44.8	
	5	0.76	2.70	64.2	
	7	0.70	2.49	67.0	
White corn	0	1.80	6.39	—	47
	1	1.75	6.21	2.9	
	3	1.33	4.72	26.1	
	5	1.15	4.08	36.1	
	7	0.90	3.20	50.0	
Barley	0	1.98	7.03	—	47
	1	1.56	5.54	21.2	
	3	1.17	4.15	40.9	
	5	0.82	2.91	58.6	
	7	0.69	2.45	65.2	
Oats	0	1.95	6.92	—	47
	1	1.60	5.68	17.4	
	3	1.49	5.29	23.6	
	5	1.30	4.62	33.3	
	7	1.31	4.65	32.8	
Rye	0	1.68	5.96	—	47
	1	1.21	4.30	28.0	
	4	0.12	0.43	92.9	
	5	0.00	0.00	100.0	
Colored ragi	0	1.72	6.10	—	91
	1	1.39	4.93	19.2	
	2	1.03	3.65	40.2	
	3	0.58	2.06	66.2	
White ragi	0	1.53	5.43	—	91
	1	1.19	4.22	22.3	
	2	0.79	2.80	48.4	
	3	0.51	1.81	66.7	

Note: All cereals were germinated at a temperature 22 to 25°C. Phytate content was calculated by assuming that it contains 28.20% phosphorus.

stages (3 days) of rye germination; however, the phytate content was reduced to 67%. Ashton and Williams[48] observed that germination of oats is accompanied by a gradual breakdown of phytate and simultaneous release of inorganic phosphorus. Complete hydrolysis of phytate occurs in sorghum after 10 days of germination.[49] The investigations of Hall and Hodges[50] indicate that phytate disappears completely from the oat endosperm during 8-day germination. Finney et al.[42] observed a reduction of 40 to 60% of phytate in several wheat varieties during five days of germination. On the other hand, Mihailovic et al.[51] noted that phytate completely disappeared in wheat by the seventh day of germination. They examined hydrolysis products at various stages of germination using circular paper chromatography and concluded that the enzymic hydrolysis of phytate occurred in a stepwise manner with the formation of intermedi-

TABLE 9
Phytate Phosphorus Hydrolysis During Fermentation and Cooking of Bakery Products[53]

Flour used	Product	Phytate originally present in flour (mg/100 g)	Amount of phytate P hydrolyzed (mg/100 g)	% Phytate P hydrolyzed
White Flour (70% extraction)	Bread made with yeast	51	43.4	85.0
National wheat meal (85% extraction)	Bread made with yeast	127	87.6	69.0
Wheat meal (92% extraction)	Bread made with yeast	214	66.3	31.0
Wheat meal (92% extraction)	Bread made with baking powder	214	10.7	5.0
Wheat meal (92% extraction)	Steamed pudding	214	34.2	16.0
Wheat meal (92% extraction)	Pastry	214	0.0	0.0
White flour with added sodium phytate	Bread made with baking powder	214	32.1	15.0
White flour with added sodium phytate	Steamed pudding	214	128.3	60.0
White flour with added sodium phytate	Pastry	214	32.1	15.0

ates — penta-, tetra-, tri-, di-, and monophosphates of myoinositol. Sobolev[52] came to similar conclusions for both *in vitro* and *in vivo* studies with castor seeds.

Overall, phytate reduction in cereals and legumes during germination may enhance the nutritional quality of these foods with respect to mineral bioavailability and protein digestibility.

IV. FERMENTATION AND BREADMAKING

Fermentation of cereals and legumes significantly reduces the phytate content due to endogenous phytase of cereals and legumes and that of added yeast and other microorganisms. Phytate hydrolysis in the bread depends on the method of bread preparation, type of flour used, amount of yeast added to the dough, and the time of dough fermentation. Widdowson[53] studied phytate destruction or hydrolysis in flour by phytase during different modes of fermentation and preparation. Some of the results are presented in Table 9. The percent phytate (85%) hydrolyzed in the bread made from 70% extraction white flour is much higher than that in the bread made from 92% extraction wheat flour. Addition of yeast to the bread preparations significantly increases phytate hydrolysis. Very little phytate is hydrolyzed from the bread, pastry, and puddings prepared without added yeast (Table 9). Pringle and Moran[54] found 59, 64, and 76% phytate hydrolysis in bread prepared with 85% extraction wheat flour and baked after 3, 5, and 8 h of fermentation. The high optimum temperature of phytase may also permit some phytate hydrolysis during initial stages of baking. Mollgaard et al.[55] observed a 39 to 49% reduction of phytate in bread made from 92% extraction rye flour and fermented with a lactic acid bacteria. Several reports have shown that there is a loss of phytate during breadmaking: 22% for whole wheat pup loaves and 66% for white pup loaves,[56] 31 to 50% for whole wheat breads,[57,58] 88 to 100% for white breads,[57-59] 13 to 39% in Iranian village flat breads,[58,60] and 17.4 to 46.2% in chapatis made with and without yeast.[61,62] Hydrolysis of phytate occurs during all stages of breadmaking.[56-58,63-65] Factors that affect the extent of phytate hydrolysis in bread during breadmaking include: (1) extraction rate of flour,[53,60,66] (2) freshness of the flour,[55] (3) the amount of yeast present,[59,66,67] (4) leavening time and temperature,[53,54,57,63,67] (5) pH of the dough,[53,55,61,63,68] (6) water content of the dough,[55] (7) presence and type of calcium salts[54,64,69] and magnesium chloride[69] added, (8) presence and concentration of oxy acids[55] and permitted additives such as vitamin C[64] and sodium bicarbonate,[53,70] and (9) solubility of phytate, especially its magnesium salt.[63] Use of sodium bicarbonate as a bread-leavening agent in bread preparation results in decreased phytate hydrolysis.

Several other additives inhibit phytate hydrolysis. Ranhotra[64] and Zemel and Shelef[69] found that calcium salts and magnesium chloride inhibit phytate hydrolysis. In contrast, McKenzie-Parnell and Davies[65] found that phytate hydrolysis is not affected in breads prepared with whole wheat flour and calcium carbonate. Iran, Pakistan, Iraq, Egypt, and other countries use sodium bicarbonate as a bread-leavening agent in the preparation of several regional breads and consumption of those breads may result in mineral deficiencies.

Ranhotra[64] investigated phytate hydrolysis during different steps of baking with white flour, blend A (containing 30% wheat protein concentrate [WPC]), and all WPC blend. The WPC was high in phytate and contained about 92% phosphorus in the form of phytate. A majority of the phytate is hydrolyzed during fermentation (storage time, 210 min) in the preparation of white bread and there is a simultaneous increase in free inorganic phosphorus (Figure 4). For blend A, there was a substantial hydrolysis of phytate and increase in inorganic phosphorus during fermentation. With all WPC blend, hydrolysis of phytate was rather slow during sponge time but greatly increased during the subsequent 30-min period (floor time), accompanied by a sharp increase in inorganic phosphorus. Ranhotra[64] observed an inverse relationship between the phytate initially present in the unbaked blend and the amount of phytate hydrolyzed during baking when white flour was increasingly replaced by WPC (Figure 5). Hydrolysis of phytate was greater in the blends with 30% or less substitution of WPC. The amount of phytate hydrolyzed progressively decreased during breadmaking as the WPC incorporation was increased above 30% in the blends. In a subsequent study, Ranhotra[74] suggested that the decreased rate of phytate hydrolysis observed in blends with increased (more than 30% incorporation) WPC during bread making was due to increased inhibition of phytase and/or rephosphorylation of partially hydrolyzed phytate. Ranhotra et al.[59] showed that all of the phytate in wheat bread and more than 3/4 of the phytate in soy-fortified wheat flour (soy 10%, wheat 90%) was hydrolyzed during the process of breadmaking apparently due to phytases in the wheat and/or yeast. Less than half of the phytate was hydrolyzed, when soy-fortified (soy 10%, wheat 90%) bread was made without added yeast (Table 10). Hydrolysis increased substantially when yeast was added, and at levels normally used in the bread formulation (9 g), phytate hydrolysis is maximum (88%). Harland and Harland[67] reported that addition of yeast and increase in length of fermentation time (rising time) both enhance phytate hydrolysis in breads prepared from rye flour, white flour, and whole wheat flour. They observed a major reduction in phytate during the first 2 h of rising and only a small decrease in next 6 h of rising for rye, white, and whole wheat breads made with either one or two packets of yeast. McKenzie-Parnell and Davies[65] found that phytate is lost at all stages of home-made brown and white bread preparation (Table 11). The highest rate of phytate hydrolysis occurs during the proving time; for brown and white breads baking and cooling further decreases phytate contents. Dagher et al.[88] reported that more than 30% of the original phytate is lost from Arabic breads made with and without wheat bran.

Reinhold[72] investigated phytate hydrolysis during yeast fermentation in sponges or doughs prepared from Iranian whole wheat meals of different extraction rates (Bazari, Sangak, and Tanok wheat meals with respective extraction rates of 75 to 85, 85 to 90, and 95 to 100%). Phytate hydrolyzed rapidly in whole meals of Bazari and Sangak with a simultaneous increase

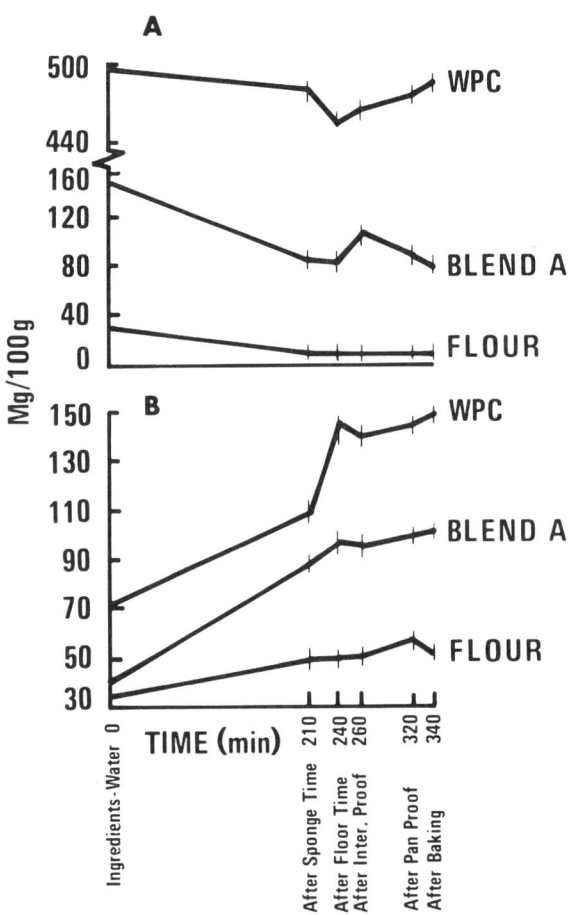

FIGURE 4. Phytic acid hydrolysis during various steps in baking; A = Phytic acid phosphorus, B = inorganic phosphorus. WPC = wheat protein concentrate, blend A = 30% WPC + 70% white flour, flour = 100% white flour. (From Ranhotra, G. S., *J. Food Sci.*, 37, 12, 1972. With permission.)

in acid-soluble phosphorus content (Figure 6). About 1/3 of the phytate disappeared within 2 h of fermentation in Bazari and Sangak doughs. The decrease of phytate in Tanok dough is much less. In Tanok dough, hindrance of phytate hydrolysis persisted after 4 h of fermentation and was still evident after 16 h (Figure 6). Reinhold[72] suggested several factors for the slower rate of phytate hydrolysis in Tanok dough: (1) presence of inhibitors such as calcium and other metal ions, which form stable salts with phytate that are resistant to phytase attack, (2) the existence of phytate in combination with certain proteins in the form of complexes in seeds, which may not be vulnerable to phytase attack, and (3) the high concentration of phytate in the Tanok whole wheat meal, which inhibits phytase action. McKenzie-Parnell and Davies[65] noticed little or no phytate hydrolysis in unleavened bran and whole meal breads (Table 12). Less than 10% of the phytate was hydrolyzed in unleavened whole meal bread. However, during the preparation of some Pakistani flat breads (chapati and roti), dalya (wheat porridge), and puri much higher amounts of phytate is hydrolyzed (Table 13).[62] From 44 to 61% and 80 to 85% of the phytate, respectively, is lost during preparation of flat breads and puri. In the preparation of dalya there is an 88% reduction of phytate.

Chompreeda and Fields[90] reported that a natural lactic acid fermentation of corn meal decreases phytate by 78%. A 9-h fermentation results in the maximum reduction (27%) of phytate (Table 14) in Rabadi— a fermented pearl millet food.[73]

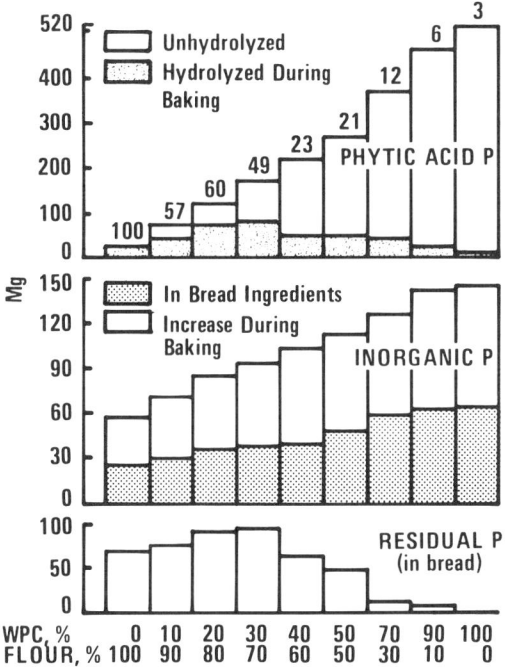

FIGURE 5. Effect of the amount of WPC (wheat protein concentrate) on hydrolysis of phytic acid during baking. Values represent the total (mg) in unbaked ingredients or the resultant pup loaf. Values for percent phytic acid hydrolyzed are shown on top of bars. (From Ranhotra, G. S., *J. Food Sci.*, 37, 12, 1972. With permission.)

TABLE 10
Effect of Added Yeast in Breadmaking on the Hydrolysis of Phytic Acid in Soy-Fortified Wheat Flour[59]

	Yeast (g/lb loaf)			
	0	3	9	15
Phytic acid P in bread ingredients representing 1 lb				
Loaf (mg/loaf)	307.2	307.2	307.2	307.2
In bread (mg/loaf)	167.0	77.2	37.2	42.4
Hydrolyzed (%)	45.6	74.7	87.9	86.2

Note: Yeast has a phytase activity of 135 units/100 g.

Hydrolysis of phytate occurs during various stages of tempeh preparation with *Rhizopus oligosporus*.[74-76] Boiling or steaming results in a reduction of phytate in soybeans (Tables 15 and 16). About 33 to 55% of the soybean phytate is hydrolyzed during tempeh fermentation with *Rhizopus*. Storage of fresh tempeh and frying both produce a further reduction in phytate content (Table 16). Less than 10% of the phytate remains in tempeh after fermentation, storage, and frying. Use of germinated (12 or 24 h) soybeans in tempeh preparation and fermentation further decreases phytate content.[77] Phytate hydrolysis occurs in tempeh due to the action of phytase which is produced during fermentation by the mold *Rhizopus oligosporus*.[74] Fardiaz and Markakis[78] obtained a 60 to 65% decrease in the phytate content of oncom (fermented peanut press cake) fermented with *Rhizopus oligosporus* or *Neurospora intermedia*. Maximum

TABLE 11
Phytate Hydrolysis During Various Stages of Leavened Homemade Brown and White Breads Preparation[65]

	Phytate	
	Content (g/100g)	% Hydrolyzed
Brown Breads		
Bread yeast ingredients	0.52	—
Mixing	0.47	9.6
Kneading	0.43	17.3
Proving (80 min)	0.32	31.9
Baking and cooling	0.22	53.2
White Breads		
Bread yeast ingredients	0.10	—
Mixing	0.10	—
Kneading	0.10	—
Proving	0.03	30.0
Baking and cooling	0.01	90.0

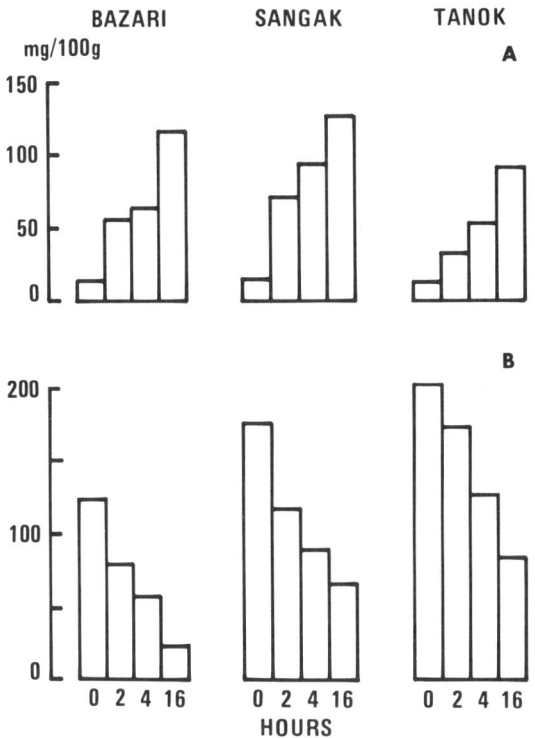

FIGURE 6. (A) Production of acid-soluble phosphorus which accounts for nearly all of the phytate phosphorus that was hydrolyed in the two whole meals of lower extraction rates, but not in that of the Tanok meal. (B) Disappearance of phytate by action of yeast in sponges of whole wheat meals of Bazari (75 to 85% extraction rate), Sangak (85 to 90% extraction rate), and Tanok (95 to 100% extraction rate). (From Reinhold, J. G., *J. Am. Diet. Assoc.*, 66, 38, 1975. With permission from The American Dietetic Association.)

TABLE 12
Phytate Loss in Unleavened Bran and Whole Meal Breads[65]

	Phytate	
	Content (g/100 g)	% Hydrolyzed
Bran Breads		
Bran bread ingredients	0.65	—
Dough making	0.64	—
Baking and cooling	0.65	—
Whole Meal Breads		
Whole meal ingredients	1.01	—
Dough mixing	0.99	2.0
Baking and cooling	0.92	8.9

TABLE 13
Changes in Phytate During Preparation of Flat Breads (Chapati and Roti), Dalya (Wheat Porridge), and Puri[62]

Ingredients	Phytate content (%)	Phytate loss (%)
Flat Breads		
Whole wheat flour	1.20	—
Dough resting (2 h)	0.86	28.3
Chapati	0.67	44.2
Roti	0.47	60.8
Dalya (Wheat Porridge)		
Whrsy	1.30	—
Roasted wheat	0.99	25.0
Dalya	0.16	87.9
Puri		
White flour (Maida)	0.20	—
Dough resting	0.18	10.0
Puri (light brown)	0.04	80.0
Puri (brown)	0.03	85.0

TABLE 14
Effect of Fermentation on Phytate Content of Rabadi (A Fermented Pearl Millet Food) at 35°C[73]

Fermentation time (h)	Phytate content (mg/100 g)	Phytate % hydrolyzed
0	480	11.3
3	426	17.9
6	394	27.1
9	350	27.1

TABLE 15
Phytate Content of Soybeans and Tempeh[74]

Sample	Phytate content (%)	Phytate hydrolyzed (%)
Soybeans, raw	1.41	—
Soybeans, soaked	1.43	—
Soybeans, boiled	1.23	14.0
Tempeh	0.96	32.9

TABLE 16
Phytate Content of Soybeans, Intermediate Products and Tempeh[76]

Sample	Phytate content (%)	Phytate hydrolyzed (%)
Whole soybeans	1.07	—
Dehulled cotyledons	1.65	—
Steamed (30 min)	1.48	10.3
Drained and Cooled	1.47	10.9
Tempeh fermented	0.75	54.5
Fried fresh tempeh	0.38	77.0
Tempeh stored (2 weeks at 5°C)	0.18	89.1
Fried stored tempeh (2 weeks at 5°C)	0.09	94.5

amounts of phytate remained in the oncom prepared with the Indonesian strain of *Neurospora intermedia*. Some reduction in phytate content occurs in black gram, rice, and black gram and rice blends during natural fermentation.[79] After 45 h of fermentation, 13.3 and 48.8% of phytate hydrolyzed in black gram, and black gram and rice blends, respectively (Table 17). In the case of rice blend, fermentation for 8 h resulted in complete hydrolysis of phytate.

V. SOAKING, AUTOLYSIS, AND OTHER PROCESSES

Soaking of cereals and legumes in water or a solution at an optimum pH and temperature results in phytase activation and phytate hydrolysis. Mellanby[47] found that phytate hydrolysis in intact whole wheat grain is slow compared to that in ground wheat when both were incubated at 45°C and pH 4.50. The intact whole wheat grain still contained about 17%(or 38 mg) of phytate at the end of 12 h (Figure 7). On the other hand, all of the phytate was hydrolyzed in ground wheat within 1 h. Differences similar to those between whole and ground wheat, relative to the rate of hydrolysis of phytate, have been reported for other cereals.[47] The variations in the rate of phytate hydrolysis observed in ground wheat compared to whole wheat may be due to the mechanical effect of the grinding. Mellanby[47] also studied enzymatic hydrolysis of phytate for other cereals. All the grains, with the exception of corn, were ground to approximately the same degree of fineness. The corn sample was coarser. Rye and wheat have the greatest phytase activity; in rye, all of the phytate is hydrolyzed in 45 min, and in wheat, the corresponding time for complete hydrolysis of phytate is 60 min (Table 18). In oats, yellow corn, and white corn, phytase acts much more slowly and even after 12 h a large part of the phytate remains in these grains. All the phytate is hydrolyzed in barley in 2 h. Glass and Geddes[80] found increased inorganic phosphorus and decreased phytate in wheat stored at increased moisture content and temperature.

TABLE 17
Effect of Fermentation on Phytate Content in Black Gram, Rice, and Black Gram and Rice Blends[79]

Fermentation time (h)	Phytate content (mg/100 g)	Phytate % hydrolyzed
Black Gram		
Raw	17.04	—
0[a]	17.04	—
4	16.90	0.8
8	16.54	2.9
12	16.54	2.9
16	16.54	3.8
20	16.40	3.8
24	14.98	12.1
45	14.77	13.3
Rice		
Raw	3.44	—
0[a]	2.31	32.9
4	0.25	92.7
8	0.00	100.0
Black Gram and Rice[b]		
0[a]	8.95	—
4	7.63	14.8
8	6.85	23.5
12	5.72	36.1
16	5.50	38.6
20	5.36	40.1
24	4.90	45.3
45	4.58	48.8

[a] Refers to samples which were soaked for 2 h before blending. Data expressed on dry weight basis.
[b] Black gram and rice blend contained equal amounts of black gram and rice.

Ferrel[81] showed that over 80% of the phytate could be hydrolyzed in hard red winter wheat by incubating the wheat for 4 h at temperatures of 35 and 55°C and pH 5.20. Dagher et al.[88] reported that about 77% of the original phytate is hydrolyzed in wheat bran when incubated at pH 5.2 and 55°C for 90 min, with the fastest rate of hydrolysis during the first 15 min. Preece et al.[82] studied the behavior of barley phytate during malting and brewing processes. They indicated that most of the phytate in barley disappears during the malting and brewing processes.

Kon et al.[83] explored ways to maximize phytate hydrolysis in California Small White (CSW) beans using different treatments. These treatments included: (1) varying the pH of the medium, (2) incubation of beans at 55°C for 20 h, (3) cooking of beans for 2 h prior to incubation, (4) addition of 1.2% wheat phytase to beans during incubation, and (5) combinations of these treatments. They were able to hydrolyze 77 and 93% phytate in two treatments and about 60 to 70% of phytate in four other treatments. These values for phytate hydrolysis were based on the inorganic phosphorus measurement. Becker et al.[84] also observed a decrease in the phytate content of a CSW bean slurry during incubation. Chang et al.[85] studied the conditions for phytase activation and the resulting effect on the phytate level in CSW beans. At 50°C, hydrolysis of phytate from the CSW beans was 31% and reached a maximum of 49% at 60°C (Table 19).

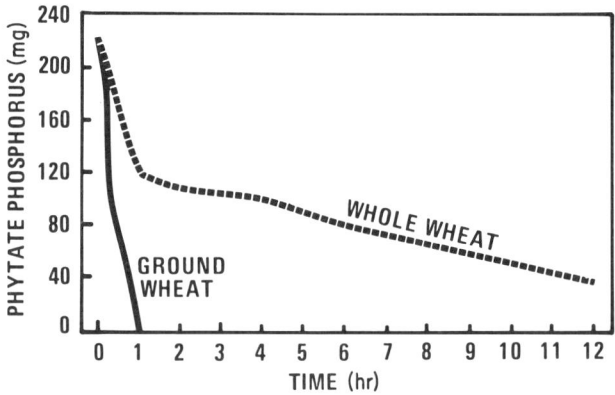

FIGURE 7. Rate of phytate hydrolysis by phytase in wheat (wheat was stirred in water at a pH of 4.50 and a temperature of 45°C). (From Mellanby, E., *A Story of Nutritional Research,* Williams and Wilkins, Baltimore, 1950, 260. With permission.)

Hydrolysis was substantially less at temperatures below 50°C or above 70°C. Further, Chang et al.[85] reported that after 10 h of incubation of CSW beans at 60°C, only a negligible amount of phytate was found in the CSW beans, approximately 75% of the total phytate being hydrolyzed, and 25% being diffused into the water in which the beans were incubated. Phytate in CSW beans was hydrolyzed throughout the pH range of 4.0 to 5.8 with maximum hydrolysis at pH 5.0 to 5.5 (56.3 to 61.5% phytate hydrolysis during 24 h of incubation) (Table 20). Hydrolysis of phytate is greatest in CSW beans when the beans are incubated in distilled water at 60°C for 4 h compared to incubation of the beans at 50°C in either distilled water or in acetate buffer for 3 h (Table 21). About 32% of the phytate is hydrolyzed in rehydrated CSW beans exposed to water-saturated air at 60°C for 6 h.

Preheat treatment at 60°C causes phytate hydrolysis in mung bean, soybean, lima bean, and wheat (Table 22). There is a 37% decrease in phytate content when whole soybeans are incubated overnight at room temperature, as compared to 1% phytate hydrolysis for soybean slurry incubated overnight at 60°C. Greatest amounts of phytate hydrolyzed are in lima beans and wheat slurries when they are incubated at 60°C. In contrast, Ferrel[81] found that the hydrolysis of phytate in beans was slow at both temperatures (35 and 55°C).

Tabekhia and Luh[21] showed decreases of 7.7, 8.1, 13.2, and 19.1% of phytate, respectively, for black-eyed, red kidney, mung, and pink beans, on soaking these beans for 12 h at 24°C in tap water. There is very little phytate hydrolysis in faba beans soaked at different temperatures (20, 35, 50 and 65°C) for varying lengths of time (8, 20, and 24 h).[16] There is a 35% decrease in the phytate content of peas soaked at 60°C for 6.5 h[24] and 46 to 50% in moth beans soaked either in water or salt solution for 12 h.[23] Iyer et al.[22] reported 52.7, 69.6, and 51.7% phytate hydrolysis, respectively, in pinto, Great Northern, and red kidney beans soaked in distilled water for 18 h at room temperature (Table 23). However, phytate hydrolysis is lower in these beans, when they were soaked in a mixed salt solution (Table 23). In contrast, Deshpande and Cheryan[86] obtained less than a 9.0% decrease in the phytate content of dry beans (pinto, sanilac, cranberry, and viva pink) soaked either in distilled water, sodium bicarbonate, or mixed salt solution for 12 h. They implied that the low phytate hydrolysis in the beans soaked in mixed salt and other solutions was due to formation of insoluble metal-phytate complexes and their subsequent resistance to phytase attack and nondiffusibility of complexes from the beans to the soaking medium. Use of water as a soaking medium appears to be effective in many cases for activation of phytase and phytate hydrolysis in certain legumes and cereals when they are incubated at a temperature of 50 to 60°C.

TABLE 18
The Rate of Phytate Hydrolysis by Phytases in Various Ground Cereals at 45°C and pH 4.50

Time of incubation	Wheat phytate Content (mg/100 g)	Hydrolyzed (%)	Rye phytate Content (mg/100 g)	Hydrolyzed (%)	Barley phytate Content (mg/100 g)	Hydrolyzed (%)	Oat phytate Content (mg/100 g)	Hydrolyzed (%)	Yellow corn phytate Content (mg/100 g)	Hydrolyzed (%)	White corn phytate Content (mg/100 g)	Hydrolyzed (%)
0 min	232	—	168	—	198	—	194	—	212	—	180	—
7 min	190	18.1	156	7.1	192	3.0	—	—	—	—	—	—
15 min	137	40.9	125	25.6	164	17.2	—	—	—	—	—	—
30 min	98	57.8	23	86.3	110	44.4	—	—	—	—	—	—
45 min	50	78.4	0	100.0	75	62.2	—	—	—	—	—	—
1 h	0	100.0	—	—	36	81.8	181	6.7	194	8.5	175	2.8
2 h	—	—	—	—	0	100.0	156	19.6	189	10.8	153	15.0
4 h	—	—	—	—	—	—	154	20.6	180	15.1	141	21.7
6 h	—	—	—	—	—	—	152	21.6	176	17.0	—	—
12 h	—	—	—	—	—	—	144	25.8	157	25.9	130	27.8

Note: Experimental conditions: 5 g of ground cereal were suspended in 25 ml of acetate buffer (pH 4.50) and shaken to the whole length of incubation at 45°C.

From Mellanby, E., *A Story of Nutritional Research*, Williams & Wilkins, Baltimore, 1950, 261. With permission.

TABLE 19
Effect of Temperature on the Autolysis of Phytate in California Small White Beans[85]

Temperature (°C)	Phytate content (mg/g)	Phytate hydrolyzed (%)
23	2.88	0.0
40	2.55	11.5
50	1.98	31.3
60	1.46	49.3
70	2.34	18.8
80	2.37	14.2
90	2.55	11.5

Note: Raw beans contained 2.88 mg/g of phytate. Experimental conditions: 5 g of beans were incubated with 50 ml of distilled water at various temperatures for 3 h.

TABLE 20
Effect of pH on Autolysis of Phytate in California Small White Beans at 50°C[85]

pH	Phytate content (mg/g)	Phytate hydrolyzed[a] (%)
4.0	1.61	44.1
4.5	1.65	42.7
5.0	1.26	56.3
5.5	1.11	61.5
5.8	1.94	32.6

Note: Raw beans contained 2.88 mg/g of phytate. Experimental conditions: 5 g of beans were incubated in 40 ml of 2.0 M acetate buffer at various pH values for 24 h.

[a] Calculated based on the raw beans phytate content.

TABLE 21
Effect of Preheat Treatment at 60°C on Hydrolysis of Phytate in California Small White Beans under Various Conditions[85]

Conditions	Phytate content (mg/g)	Phytate hydrolyzed (%)
Dry beans soaked overnight at room temperature	2.88	0.0
Dry beans incubated with 10 volumes of distilled water at 60°C for 4 h	1.16	59.7
Rehydrated beans[a] exposed to 60°C water saturated air for 6 h	1.97	31.6
Dry beans incubated in distilled water at 60°C for 1 h, then transferred to 50°C water or to pH 5.0 buffer solution for 3 h	1.63	43.4
Rehydrated beans[a] frozen overnight, thawed at room temperature, and then exposed to 60°C water saturated air for 6 h	1.45	49.7

[a] Beans were soaked in distilled water at room temperature for about 16 h.

TABLE 22
Effects of Soaking and Heating on Hydrolysis of Phytate in Legumes and Wheat[85]

Conditions	Mung bean phytate		Soybean phytate		Lima bean phytate		Wheat phytate	
	Content (mg/g)	Hydrolyzed (%)	Content (mg/g)	Hydrolyzed (%)	Content (mg/g)	Hydrolyzed (%)	Content (mg/g)	Hydrolyzed (%)
Raw flour	2.21	—	3.06	—	2.30	—	2.36	—
Whole seeds or grains soaked in water for overnight at room temperature	2.16	2.3	1.93	36.9	2.46	0.0	2.26	4.2
Slurry incubated in water for overnight at 60°C	1.26	42.9	3.03	1.0	0.48	80.5	0.85	64.0
Whole seeds or grains incubated in water for 5 h at 60°C	—	—	—	—	1.36	44.7	1.81	23.3
Rehydrated whole seeds or grains incubated in water saturated air for overnight at 60°C	2.38	37.6	2.51	18.0	0.84	65.9	1.72	27.1

Note: Experimental conditions: 5 g of legumes except lima beans or wheat were used. In case of lima beans 10 g sample was used. For wet incubation 10 volumes of distilled water was used as medium.

TABLE 23
Effects of Soaking on the Phytate Content of Great Northern, Kidney, and Pinto Beans[22]

Treatment	Great Northern bean phytate		Kidney bean phytate		Pinto bean phytate	
	Content (mg/g)	Hydrolyzed (%)	Content (mg/g)	Hydrolyzed (%)	Content (mg/g)	Hydrolyzed (%)
Raw beans	4.6	—	5.8	—	5.5	—
Soaked in distilled water for 18 h at room temperature	1.4	69.6	2.8	51.7	2.6	52.7
Soaking in mixed salt solution[a] for 18 h at room temperature, pH 7.0	4.2	8.7	4.5	22.4	4.4	18.2

[a] Mixed salt solution is prepared to contain 2.5% sodium chloride, 1.5% sodium bicarbonate, 0.5% sodium carbonate and 1% sodium tripolyphosphate in distilled water.

REFERENCES

1. **Deshpande, S. S., Sathe, S. K., and Salunkhe, D. K.,** Dry beans of *Phaseolus:* a review. III, *CRC Crit. Rev. Nutr. Food Sci.,* 21, 137, 1984.
2. **Moscoso, W., Bourne, M. C., and Hood, L. F.,** Relationship between the hard-to-cook phenomenon in red kidney beans and water absorption, puncture force, pectin, phytic acid, and minerals, *J. Food Sci.,* 49, 1577, 1984.
3. **Mattson, S.,** The cookability of yellow peas: a colloid chemical and biochemical study, *Acta Agric. Suecana II,* 2, 185, 1946.
4. **Mattson, S., Akerberg, E., Erikson, E., Koulter-Anderson, E., and Vahtras, K.,** Factors determining the composition and cookability of peas, *Acta Agric. Scandinavia,* 1, 40, 1950.
5. **Muller, F. M.,** Cooking quality of pulses, *J. Sci. Food Agric.,* 18, 292, 1967.
6. **Rosenbaum, T. M., Henneberry, G. O., and Baker, B. E.,** Constitution of leguminous seeds. VI Ease of cooking field peas *(Pisum sativum)* in relation to phytic acid content and calcium diffusion, *J. Sci. Food Agric.,* 17, 237, 1966.
7. **Kumar, K. G., Venkataraman, L. V., Jaya, T. V., and Krishnamurthy, K. S.,** Cooking characteristics of some germinated legumes: changes in phytins, Ca^{++}, Mg^{++}, and pectins, *J. Food Sci.,* 43, 85, 1978.
8. **Kon, S.,** Effect of soaking temperature on cooking and nutritional quality of beans, *J. Food Sci.,* 44, 1329, 1979.
9. **Kon, S. and Sanshuck, D. W.,** Phytate content and its effect on cooking quality of beans, *J. Food Process. Preserv.,* 5, 169, 1981.
10. **Jones, P. M. B. and Boulter, D.,** The cause of reduced cooking rate in *Phaseolus vulgaris* following adverse storage conditions, *J. Food Sci.,* 48, 623, 1983.
11. **Shehata, A. M., Abu-Bakr, T. M., and El-Shimi, N. M.,** Phytate, phosphorus and calcium contents of mature seeds of *Vicia faba* L. and their relation to texture of pressure-cooked faba beans, *J. Food Process. Preser.,* 7, 185, 1983.
12. **Sievwright, C. A. and Shipe, W. F.,** Effect of storage conditions and chemical treatments on firmness, *in vitro* protein digestibility, condensed tannins, phytic acid, and divalent cations of cooked black beans *(Phaseolus vulgaris* L.), *J. Food Sci.,* 51, 982, 1986.
13. **Hincks, M. J. and Stanley, D. W.,** Multiple mechanisms of bean hardening, *J. Food Technol.,* 21, 731, 1986.
14. **Crean, D. E. C. and Haisman, D. R.,** The interaction between phytic acid and divalent cations during the cooking of dried peas, *J. Sci. Food Agric.,* 14, 824, 1963.
15. **Rosenbaum, T. M. and Baker, B. E.,** Constitution of leguminous seeds VII.Ease of cooking field peas *(Pisum sativum)* in relation to phytic acid content and calcium diffusion, *J. Sci. Food Agric.,* 20, 709, 1969.
16. **Henderson, H. M. and Ankrah, S. A.,** The relationship of endogenous phytase, phytate, and moisture uptake with cooking time in *Vicia faba* minor cv. Aladin, *Food Chem.,* 17, 1, 1985.

17. **Proctor, J. P. and Watts, B. M.,** Effect of cultivar, growing location, moisture, and phytate content on the cooking times of freshly harvested navy beans, *Can. J. Plant Sci.,* 67, 923, 1987.
18. **Reddy, N. R., Balakrishnan, C. V., and Salunkhe, D. K.,** Phytate phosphorus and mineral changes during germination and cooking of black gram *(Phaseolus mungo L.)* seeds, *J. Food Sci.,* 43, 540, 1978.
19. **Reddy, N. R. and Salunkhe, D. K.,** Interactions between phytate, protein, and minerals in whey fractions of black gram, *J. Food Sci.,* 46, 564, 1981.
20. **Anon.,** Nutritional Chemistry, in *Asian Vegetable Research and Development Center — Mung Bean Report,* Shanhua, Taiwan, 1976, 28.
21. **Tabekhia, M. M. and Luh, B. S.,** Effect of germination, cooking, and canning on phosphorus and phytate retention in dry beans, *J. Food Sci.,* 45, 406, 1980.
22. **Iyer, V., Salunkhe, D. K., Sathe, S. K., and Rockland, L. B.,** Quick-cooking beans *(Phaselous vulgaris* L.) II. Phytates, oligosaccharides and antienzymes, *Qual. Plant. Plant Foods Human Nutr.,* 30, 45, 1980.
23. **Khokhar, S. and Chauhan, B. M.,** Antinutritional factors in moth bean *(Vigna aconitifolia):*varietal differences and effects of methods of domestic processing and cooking, *J. Food Sci.,* 51, 591, 1986.
24. **Beal, L. and Mehta, T.,** Zinc and phytate distribution in peas: influence of heat treatment, germination, pH, substrate and phosphorus on pea phytate and phytase, *J. Food Sci.,* 50, 96, 1985.
25. **Ologhobo, A. D. and Fetuga, B. L.,** Distribution of phosphorus and phytate on some Nigerian varieties of legumes and some effects of processing, *J. Food Sci.,* 49, 199, 1984.
26. **Clydesdale, F. M. and Camire, A. L.,** Effect of pH and heat on the binding of iron, calcium, magnesium, and zinc and the loss of phytic acid in soy flour, *J. Food Sci.,* 48, 1272, 1983.
27. **Lease, J. G.,** The effect of autoclaving sesame meal on its phytic acid content and on the availability of its zinc to the chick, *Poultry Sci.,* 45, 237, 1966.
28. **deBoland, A., Garner, G. B., and O'Dell, B. L.,** Identification and properties of phytate in certain grains and oilseed products, *J. Agric. Food Chem.,* 23, 1186, 1975.
29. **Camire, A. L. and Clydesdale, F. M.,** Interactions of soluble iron with wheat bran, *J. Food Sci.,* 47, 1296, 1982.
30. **Tangendjaja, B., Buckle, K. A., and Wootton, M.,** Dephosphorylation of phytic acid in rice bran, *J. Food Sci.,* 46, 1021, 1983.
31. **Toma, R. B. and Tabekhia, M. M.,** Changes in mineral elements and phytic acid contents during cooking of three California rice varieties, *J. Food Sci.,* 44, 619, 1979.
32. **Long, C.,** Phytase, in *Biochemists' Handbook,* Van Nostrand-Reinhold, Princeton, 1961, 259.
33. **Belavady, B. and Banerjee, S.,** Studies on the effect of germination on the phosphorus values of some common Indian pulses, *Food Res.,* 18, 223, 1953.
34. **Peers, G. F.,** The phytase of wheat, *Biochem. J.,* 53, 102, 1953.
35. **Cosgrove, D. J.,** *Inositol Phosphates: Their Chemistry, Biochemistry and Physiology,* Elsevier, New York, 1980, 157.
36. **Chang, C. W.,** Study of phytase and fluoride effects in germinating corn seeds, *Cereal Chem.,* 44, 129, 1967.
37. **Mandal, N. C. and Biswas, B. B.,** Metabolism of inositol phosphates. I. Phytase synthesis during germination in cotyledons of mung beans, *Phaseolus aureus, Plant Physiol.,* 45, 4, 1970.
38. **Walker, K. A.,** Changes in phytic acid and phytase during early development of *Phaseolus vulgaris, Planta,* 116, 91, 1974.
39. **Chen, L. H. and Pan, S. H.,** Decrease of phytates during germination of pea seeds *(Pisum sativum* L.), *Nutr. Rep. Intern.,* 46, 125, 1977.
40. **Lolas, G. M. and Markakis, P.,** The phytase of navy bean *(Phaselous vulgaris), J. Food Sci.,* 42, 1094, 1977.
41. **Kuvaea, E. B. and Kretovich, V. L.,** Phytase of germinating pea seeds, *Soviet Plant Physiol. (Engl. transl.),* 25, 290, 1978.
42. **Finney, P. L., Mason, W. R., Jeffers, H. C., El-Samahy, S. K., and Vigue, G. T.,** Effects of germination on some physical, chemical, and breadmaking properties of 12 U.S. wheat variety composites, *66th Annual Meeting, American Association of Cereal Chemists* (Abstr.), Denver, 1981.
43. **Eskin, N. A. M. and Wiebe, S.,** Changes in phytase activity and phytate during germination of two faba bean cultivars, *J. Food Sci.,* 48, 270, 1983.
44. **Sathe, S. K., Despande, S. S., Reddy, N. R., Goll, D. E., and Salunkhe, D. K.,** Effects of germination on proteins, raffinose oligosaccharides and antinutritional factors in the Great Northern beans (*Phaseolus vulgaris* L.), *J. Food Sci.,* 48, 1796, 1983.
45. **Gad, S. S., Mohamed, M. S., El-Zalaki, M. E., and Mohasseb, S. Z.,** Effect of processing on phosphorus and phytic acid contents of some Egyptian varieties of legumes, *Food Chem.,* 8, 11, 1982.
46. **Mandal, N. C., Burman, S., and Biswas, B. B.,** Isolation, purification, and characterization of phytase from germinating mung bean, *Phytochemistry,* 11, 495, 1972.
47. **Mellanby, E.,** *A Story of Nutritional Research,* Williams and Wilkins, Baltimore, 1950, 263.
48. **Ashton, W. M. and Williams, P. C.,** The phosphorus compounds of oats. I. The content of phytate phosphorus, *J. Sci. Food Agric.,* 9, 505, 1958.

49. **Glennie, C. W., Cilliers, J. J. L., and Geyer, H. L.,** Changes in phytate and related compounds in grain sorghum during germination, *Nutr. Rep. Intern.,* 32, 349, 1985.
50. **Hall, J. R. and Hodges, T. K.,** Phosphorus metabolism of germinating oat seeds, *Plant Physiol.,* 41, 1459, 1966.
51. **Mihailovic, M. L., Antic, M., and Hadzijev, D.,** Chemical investigation of wheat. VIII. Dynamics of various forms of phosphorus in wheat during its ontogenesis. The extent and mechanism of phytic acid decomposition in germinating wheat grain, *Plant Soil.,* 23, 117, 1965.
52. **Sobolev, A. M.,** Enzymatic hydrolysis of phytin *in vivo* and in germinating seeds, *Soviet Plant Physiol.,* (Engl. transl.), 9, 263, 1963.
53. **Widdowson, E. M.,** Phytic acid and the preparation of food, *Nature (London),* 148, 219, 1941.
54. **Pringle, W. J. S. and Moran, T.,** Phytic acid and its destruction in baking, *J. Soc. Chem. Ind. (London),* 61, 108, 1942.
55. **Mollgaard, H., Lorenzen, K., Hansen, I. G., and Christensen, P. E.,** On phytic acid, its importance in metabolism and its enzymatic cleavage in bread supplemented with calcium, *Biochem. J.,* 40, 589, 1946.
56. **Tangkongchitr, U., Seib, P. A., and Hoseny, R. C.,** Phytic acid. II. Its fate during breadmaking, *Cereal Chem.,* 58, 229, 1981.
57. **Daniels, D. G. H. and Fisher, N.,** Hydrolysis of the phytate of wheat flour during breadmaking, *Br. J. Nutr.,* 46, 1, 1981.
58. **Nayini, N. R. and Markakis, P.,** Effect of fermentation time on the inositol phosphates of bread, *J. Food Sci.,* 48, 262, 1983.
59. **Ranhotra, G. S., Loewe, R. J., and Puyat, L. V.,** Phytic acid in soy and its hydrolysis during breadmaking, *J. Food Sci.,* 39, 1023, 1974.
60. **Ter-Sarkissian, N., Azar, M., Ghavifekr, H., Ferguson, T., and Hedayat, H.,** High phytic acid in Iranian breads, *J. Am. Diet. Assoc.,* 65, 651, 1974.
61. **Swaranjeet, K., Maninder, K., and Bains, G. S.,** Chapaties with leavening and supplements: changes in texture, residual sugars, and phytic acid phosphorus, *Cereal Chem.,* 59, 367, 1982.
62. **Khan, N., Zaman, R., and Elahi, M.,** Effect of processing on the phytic acid content of wheat products, *J. Agric. Food Chem.,* 34, 1010, 1986.
63. **Tangkongchitr, U., Seib, P. A., and Hoseney, R. C.,** Phytic acid. III. Two barriers to the loss of phytate during breadmaking, *Cereal Chem.,* 59, 216, 1982.
64. **Ranhotra, G. S.,** Hydrolysis during breadmaking of phytic acid in wheat protein concentrate, *J. Food Sci.,* 37, 12, 1972.
65. **McKenzie-Parnell, M. M., and Davies, N. T.,** Destruction of phytic acid during home breadmaking, *Food Chem.,* 22, 181, 1986.
66. **Reinhold, J. G., Parsa, A., Karimian, N., Hammick, J. W., and Ismail-Beigi, F.,** Availability of zinc in leavened and unleavened whole meal breads as measured by solubility, and uptake by rat intestine, *in vitro, J. Nutr.,* 104, 976, 1974.
67. **Harland, B. F. and Harland, J.,** Fermentative reduction of phytate in rye, white, and whole wheat breads, *Cereal Chem.,* 57, 226, 1980.
68. **Knorr, D., Watkins, T. R., and Carlson, B. L.,** Enzymatic reduction of phytate in whole wheat breads, *J. Food Sci.,* 46, 1866, 1981.
69. **Zemel, M. B. and Shelef, L. A.,** Phytic acid hydrolysis and soluble zinc and iron in whole wheat bread as affected by calcium containing additives, *J. Food Sci.,* 47, 535, 1982.
70. **Faridi, H. A., Finney, P. L., and Rubenthaler, G. L.,** Effect of soda leavening on phytic acid content and physical characteristics of middle eastern breads, *J. Food Sci.,* 48, 1654, 1983.
71. **Ranhotra, G. S.,** Factors affecting hydrolysis during breadmaking of phytic acid in wheat protein concentrate, *Cereal Chem.,* 50, 355, 1973.
72. **Reinhold, J. G.,** Phytate destruction by yeast fermentation in whole wheat meals, *J. Am. Diet. Assoc.,* 66, 38, 1975.
73. **Dhankher, N. and Chauhan, B. M.,** Effect of temperature and fermentation time on phytic acid and polyphenol content of Rabadi— a fermented pearl millet food, *J. Food Sci.,* 52, 828, 1987.
74. **Sudarmadji, S. and Markakis, P.,** The phytate and phytase of soybean tempeh, *J. Sci. Food Agric.,* 28, 381, 1977.
75. **Sutardi and Buckle, K. A.,** Phytic acid changes in soybeans fermented by traditional inoculum and six strains of *Rhizopus oligosporus, J. Appl., Bacteriol.,* 58, 539, 1985.
76. **Sutardi and Buckle, K. A.,** Reduction in phytic acid levels in soybeans during tempeh production, storage, and frying, *J. Food Sci.,* 50, 260, 1985.
77. **Suparmo and Markakis, P.,** Tempeh prepared from germinated soybeans, *J. Food Sci.,* 52, 1736, 1987.
78. **Fardiaz, D. and Markakis, P.,** Degradatioon of phytic acid in oncom (fermented peanut press cake), *J. Food Sci.,* 46, 523, 1981.
79. **Reddy, N. R. and Salunkhe, D. K.,** Effects of fermentation on phytate phosphorus and minerals of black gram, rice, and black gram and rice blends, *J. Food Sci.,* 45, 1708, 1980.

80. **Glass, R. L. and Geddes, W. F.,** Grain storage studies. XXVII. Inorganic phosphorus content of deteriorating wheat, *Cereal Chem.,* 36, 86, 1959.
81. **Ferrel, R. E.,** Distribution of bean and wheat inositol phosphate esters during autolysis and germination, *J. Food Sci.,* 43, 563, 1978.
82. **Preece, I. A., Gray, H. J., and Wadham, A. T.,** Studies on phytin. I. Inositol phosphates, *J. Inst. Brew.,* 66, 487, 1960.
83. **Kon, S., Olson, A. C., Frederick, D. P., Eggling, S. B., and Wagner, J. R.,** Effect of different treatments on phytate and soluble sugars in California small white beans, *J. Food Sci.,* 39, 215, 1973.
84. **Becker, R., Olson, A. C., Frederick, D. P., Kon, S., Gumbmann, M. R., and Wagner, J. R.,** Conditions for autolysis of alpha-galactosides and phytic acid in California small white beans, *J. Food Sci.,* 39, 766, 1974.
85. **Chang, R., Schwimmer, S., and Burr, H. K.,** Phytate: removal from whole dry beans by enzymatic hydrolysis and diffusion, *J. Food Sci.,* 42, 1098, 1977.
86. **Deshpande, S. S. and Cheryan, M.,** Changes in phytic acid, tannins, and trypsin inhibitory activity on soaking of dry beans (*Phaseolus vulgaris* L.), *Nutr. Rep. Intern.,* 27, 371, 1983.
87. **Longe, O. G.,** Varietal differences in chemical characteristics related to cooking quality of cowpea, *J. Food Process Preservat.,* 7, 143, 1983.
88. **Dagher, S. M., Shadarevian, S., and Birbari, W.,** Preparation of high bran Arabic bread with low phytic acid content, *J. Food Sci.,* 52, 1600, 1987.
89. **Fretzdorff, B. and Weipert, D.,** Phytic acid in cereals. I. Phytic acid and phytase in rye and rye products, *Z. Lebensm. Unters. Forsch.,* 182, 287, 1986.
90. **Chompreeda, P. T. and Fields, M. L.,** Effects of heat and fermentation on the extrability of minerals from soybean meal and corn meal blends, *J. Food Sci.,* 49, 566, 1984.
91. **Rao, P. U. and Deosthale, Y. G.,** *In vitro* availability of iron and zinc in white and colored ragi *(Eleusine coracana):* role of tannin and phytate, *Plant Food Hum. Nutr.,* 38, 35, 1988.

Chapter 11

TECHNOLOGY OF PHYTATE REMOVAL

I. INTRODUCTION

The processing methods discussed in an earlier chapter may not completely eliminate phytate in foods made from cereals or legumes. However, in some food products, certain processing methods remove some phytates resulting in improved nutritional quality and nutrient bioavailability.[1-6] Processing of whole cereal grains and legumes through physical and chemical treatments to produce products such as flour, bran, germ, starch, and protein concentrates and isolates may result in removal of large amounts of phytate. Factors such as processing conditions, extracting agents, and the form of phytate and phytate localization have been reported to influence removal of phytate from cereals and legumes.[7-13] The removal of phytate from cereals and legumes and their products by physical and chemical treatments such as mechanical processes, selective extraction, membrane filtration, ion exchange methods, and enzymatic treatment methods have been studied.[3,5,14] In many instances, a great reduction but not a complete elimination of phytate can be achieved.

II. MECHANICAL PROCESSES

In cereals, phytate is associated with specific components and can be selectively removed by mechanical processes such as milling. For example, about 89% of the phytate in corn is concentrated in the germ portion and can be removed by milling followed by germ separation.[7] In wheat, rice, triticale, and rye, phytate is concentrated in the outer layers (bran) and hence normal milling (which involves bran removal) should remove appreciable amounts of phytate from these cereals.[15-17] The degree of milling would also result in reduction of phytate in wheat flours.[18] Polishing the rice has been shown to remove significant amounts of phytate.[17,19]

Phytate in dry beans is associated with the protein bodies in the cotyledons and cannot be preferentially removed by milling. Deshpande et al.[8] reported that dehulling (removal of seed coat) significantly increases the phytate content of beans. Others[20,21] air classified legume flours in to protein and starch fractions with the intention of decreasing the content of phytate. However, air classification of legume flours did not result in a decrease in phytate in these fractions. The proportion of phytate in the protein fractions of several legume flours was nearly three times that of the original flours. The starch fractions contained decreased amounts of phytate. Legume flours require extensive processing if the protein fractions are to be used in foods.[20]

III. SELECTIVE EXTRACTION AND DIFFERENTIAL SOLUBILITY METHODS

Selective extraction and differential solubility methods to precipitate and remove phytate from soybeans have been extensively studied. Such methods usually involve extraction with either water, alkali, or salt, followed by pH adjustment so that the phytate can be removed by centrifugation or filtration. McKinney et al.[22] removed about 80% of the phytate from an alkaline soybean extract by precipitation of phytate with calcium and/or barium ions and subsequent centrifugation (Table 1). They recovered about 80% of the soybean meal nitrogen during the process of protein concentration and simultaneously removed phytate. Hartman[2] described a process for preparation of a low phytate soy protein isolate using high pH adjustment (Figure 1). The process included:

TABLE 1
Precipitation and Removal of Phytate from Soybean Meal Extract with Calcium and Barium Ions[22]

Extractant	Extract pH	Special treatment	Phosphorus in protein (%)
0.2% NaOH	9.7	0.2% Ba(OH)$_2$ added, pH 10.9; precipitate removed by centrifugation	0.46
0.2% NaOH	9.7	0.2% Ba(OH)$_2$ added; adjusted to pH 9.5 and heated to 80°C; centrifuged	0.27
0.1% NaOH	9.5	5% BaCl$_2$ added, held at 5°C for 17 h, centrifuged, Na$_2$SO$_3$ added to remove Ba; clarified and precipitated	0.18
0.1% NaOH	9.5	5% BaCl$_2$ added; held at 5°C for 17 h; centrifuged and dialyzed	0.32
0.1% Ca(OH)$_2$	9.2	0.7% NaOH added; held at 5°C, pH 11.5 for 17 h, centrifuged and dialyzed	0.20
1.0% Ca(OH)$_2$	9.2	Heated to 85°C, centrifuged, discard precipitate	0.65
1.0% Ba(OH)$_2$	—	Set extract for 2 h; centrifuged	0.20

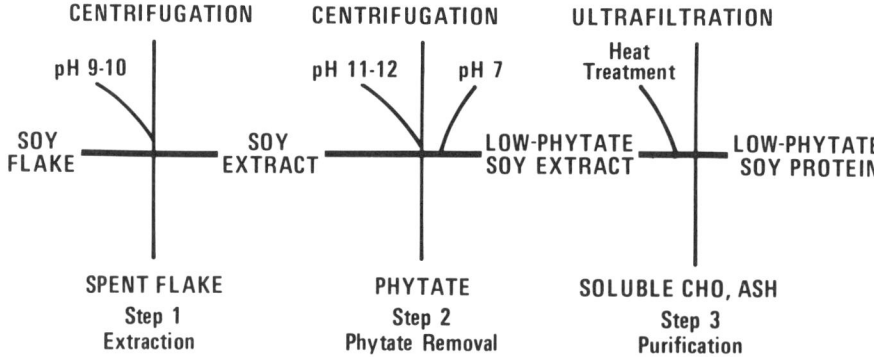

FIGURE 1. Process outline for phytic acid removal from soy protein. (From Hartman, G. H., *J. Am. Oil Chem. Soc.*, 56, 731, 1979. With permission.)

Step 1 — Soy flakes were extracted with water at high pH (9 to 10) with a water-to-flakes ratio of 16:1 and the slurry was centrifuged.

Step 2 — The phytate in the supernatant from Step 1 was precipitated by adjusting the pH of the supernatant to 11 to 12. The precipitated phytate was then removed by either centrifugation or vacuum filtration and the pH of soy extract (after removing phytate) was rapidly adjusted to 7.0 with dilute hydrochloric acid. The temperature during processing was controlled between 20 and 30°C.

Step 3 — The soy extract was then subjected to ultrafiltration to obtain a low phytate soy protein isolate.

The temperature and pH were both critical for efficient removal of phytate and prevention of protein degradation. The low phytate soy protein isolate prepared by Hartman's process contains 93% of the original soy protein and retains only 4% of the original phytate.

de Rham and Jost[1] developed three processes for preparing low phytate soy protein products from soy flour. The main features of these processes as compared to the classical process for soy protein isolate production are summarized in Table 2. In one process involving a pH 11.50

TABLE 2
Protein and Phytate Content of Soy Protein Isolates Produced by Various Processes[1]

Process	Description
pH 4.50, classical	Extraction of the defatted soy flour with water at pH 8.20, centrifugation, washing of the residue, acidification to pH 4.50, washing of the precipitate and lyophilization.
pH 5.50	Identical to pH 4.50, except the precipitation at pH 5.50, and the omission of the washing steps.
pH 11.50/5.50	Identical to pH 4.50, except on extraction at pH 11.50 with NaOH, precipitation at pH 5.50, and the omission of washing steps.
NaCl/UF	Extraction of the defatted soy flour with 10% NaCl solution, centrifugation, (first supernatant); separate filtration and ultrafiltration-diafiltration of the first supernatant and lyophilization.

	Protein (%)	Phytate content (%)	Phytate removed (%)
Soybeans	50.0	1.50	—
pH 4.50	86.0	2.84	—
pH 5.50	93.0	0.60	60.0
pH 11.50/5.50	90.0	0.18	88.0
NaCl/UF	84.0	0.14	90.7

Note: UF = ultrafiltration.

extraction, a protein isolate with 0.18% phytate (a reduction of 88%) and 90% protein was obtained. At highly alkaline pH values formation of lysinoalanine may occur, which is nutritionally undesirable; however, this substance was not detected in protein isolates prepared in the pH 11.50 process.[2,23] A soy protein isolate prepared by precipitating proteins at pH 5.50 with no salt addition contained 0.6% phytate (a reduction of 60%) and 93% protein (Table 2). Further, de Rham and Jost[1] evaluated the composition and nutritional value of protein isolates prepared by the three different processes and found that an isolate prepared by the pH 5.50 process had a protein efficiency ratio (PER) that was better than the PER of other protein isolates.

A significant difference in solubility of the soybean proteins also occurs around pH 5.50.[14] Using this principle, Ford et al.[24] developed a process to prepare soy protein isolate. They used either a combination of low pH (3.5 to 4.0) and high calcium concentration (0.04 M) or a high pH (5.0 to 5.5) and low calcium concentration (0.0025 M) to remove 90% or more of phytate from soybean extracts (Figure 2). The high pH and low calcium process resulted in a high (90%) zinc recovery in the curd as well as high recoveries of protein (90 to 93%), fat (98 to 100%), and iron (94 to 96%). Baker et al.[25] prepared soy protein concentrates with low phytate by precipitating the proteins at pH 5.50. At pH 5.50 soy proteins are insoluble whereas phytate remains in the solution. The insoluble, proteins could then be removed by centrifugation to yield low phytate protein concentrate (Table 3). Several other researchers[24,26,27] removed phytate from soy extracts by using a low pH in combination with excess calcium ions during preparation of protein isolates or concentrates. At acidic pH values both cationic protein groups and calcium ions compete for phytate binding. The higher affinity of calcium ions for phytate prevents the formation of phytate-protein complexes, thus resulting in the production of low phytate protein isolates.[14] However, the low pH with excess calcium ions process may be undesirable because at this pH some proteins will dissociate into subunits and denature.[27] This may result in off-flavored products.[28] Brooks and Morr[29] attempted to remove phytate from an alkaline soybean extract by precipitating phytate with divalent cations (calcium, barium, and zinc). They were not successful and based on the toxicity, possible off-flavor and reduced functionality properties,

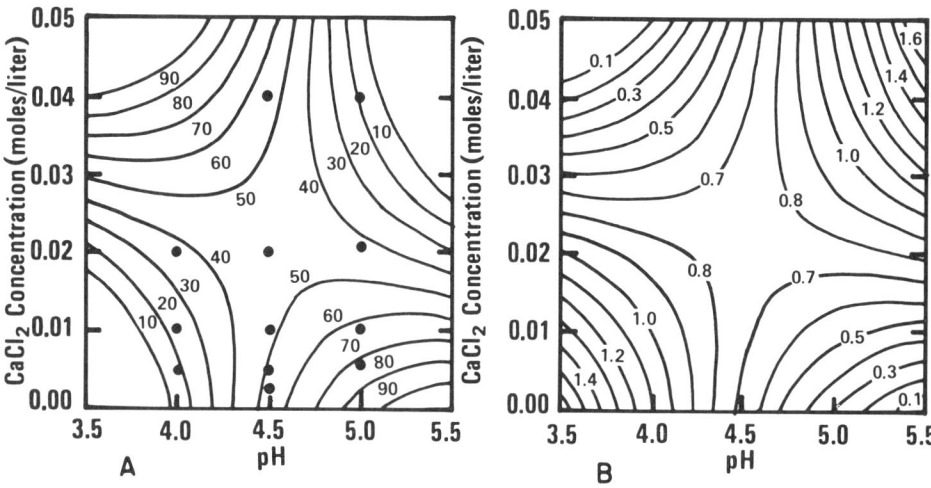

FIGURE 2. (A) Percent reduction of phytic acid from the curd as a function of pH and calcium chloride concentration (Percent reduction = 100% recovery in curd). (B) Concentration ratio of phytic acid as function of pH and calcium chloride concentration. (From Ford, J. R., Mustakas, G. C. and Schmutz, R. D., *J. Am. Oil Chem. Soc.*, 55, 371, 1978. With permission.)

TABLE 3
Composition of Soy Protein Concentrate Prepared from Soy Flour[25]

Component (%)	Soy flour	Protein product precipitated at pH 4.50	Protein product precipitated at pH 5.50
Moisture	3.30	5.70	5.80
Protein	39.00	48.80	48.50
Ash	4.70	3.30	3.30
Crude fiber	2.70	3.70	3.10
Phytate	1.30	0.73	0.29
Phytate removed	—	43.80	77.70

Note: Data expressed on dry weight basis.

concluded that the divalent cation-fractionation process is not a method for producing phytate-reduced soy isolates for food use.

A process consisting of alkaline extraction and precipitation of proteins at an acidic pH appears to reduce the phytate content in products such as protein concentrates or isolates of other legumes. Deshpande and Cheryan[30] prepared protein concentrates from five dry beans (small white, Great Northern, viva pink, cranberry, and light red kidney) by a wet extraction process involving alkaline extraction and precipitation of proteins at pH 4.50. The phytate content of bean protein concentrates ranged from 0.47 to 1.52%, representing a reduction of approximately 38 to 60% as compared to whole beans (Table 4). They attributed the presence of appreciable amounts of phytate in bean protein concentrates to its complexation with bean proteins. Recently, Thompson[46] used acylation techniques to separate proteins from phytate in navy bean flour and to produce low phytate protein isolates. She achieved the separation of phytate from the proteins in the acylated extract of navy bean flour by isoelectric precipitation at pH values 3.5 and 4.0. At these pH values, more than 88% of the proteins and less than 10% of phytate were precipitated. The low phytate protein isolates prepared by acylation techniques from navy bean flour had excellent functional properties. However, the protein isolates manufactured by

TABLE 4
Phytate Content of Whole Beans and Protein Concentrates[30]

Cultivar	Phytate (mg/g)		% Reduction in phytate
	Whole bean	Protein concentrate	
Small White	11.6	4.7	59.5
Great Northern	20.4	12.7	37.7
Viva Pink	21.6	11.4	47.2
Cranberry	26.3	13.9	47.1
Light red kidney	26.3	15.2	42.2

TABLE 5
Removal of Phosphorus and Nitrogen from Soybean Water Extract by Dialysis and Anion Exchange[31]

Treatment	Phosphorus removed (%)	Nitrogen removed (%)
Dialysis 24 h, pH 7.20	40.5	7.5
Dialysis 48 h, pH 6.50	72.0	7.0
Dialysis 48 h, pH 8.70	52.0	28.0
Anion exchange,[a] pH 7.50	56.5	4.0
Anion exchange, pH 7.00	65.0	7.5
Anion exchange, pH 6.50	82.3	24.0
Dialysis + anion exchange,[b] pH 7.00	78.2	10.1

[a] Dowex-1-X10®.
[b] Dialysis at pH 7.20 (24 h) followed by Dowex-1-X10® at pH 7.00.

acylation techniques are not commercially produced for human food due to limited data on their safety and toxicity.[46]

IV. MEMBRANE SEPARATION

Phytate can be separated from proteins and other components in legume and cereal extracts by using methods based on molecular size differences if the phytic acid or its salts are in a soluble form and not complexed with proteins in the extracts. The pH and cation concentration both play a role in the removal of phytate by dialysis. Smith and Rackis[31] attempted to remove the free phytic acid and its salts at different pH ranges by a combination of methods involving dialysis and anion-exchange resin. They reported that about 72% of the phytate could be removed by dialysis at pH 6.50 without appreciable loss of proteins (Table 5). McKinney et al.[22] found that phytate could be removed by dialyzing wet, isoelectrically precipitated curd of soybeans against a $1N$ sodium chloride solution. Extraction of the wet curd of soybeans with a saturated solution of ammonium sulfate followed by dialysis to remove the salt was also reported to be effective in reducing phytate content. McKinney and Solars[32] prepared a low phytate soy protein isolate by extracting soy proteins with sulfurous acid at pH 2.30 followed by isoelectric precipitation (using NaOH) and dialysis. The protein isolate thus prepared had 1.7 to 3.0% phytate, which could not be removed by dialysis in acid conditions due to a strong phytate-protein complexation at the low pH used for protein extraction.

Serraino and Thompson[33] found that dialysis (spectrapor membrane tubing molecular weight cut off 6,000 to 8,000) of rapeseed protein concentrate adjusted to pH 3.5 and rapeseed flour

TABLE 6
Effects of pH and Dialysis on Removal of Protein, Phytate, and Minerals from Great Northern Bean Flour and Combined Density Fraction[13]

Component	Undialyzed material	% of the original remaining at			
		pH 4.0	pH 6.3	pH 7.0	pH 9.0
Great Northern Bean Flour					
Yield	5.0 g	76.0	76.0	76.0	76.0
Protein	25.7 %	82.5	85.2	84.3	91.7
Phytate	26.6 mg/g	46.3	59.7	42.3	56.0
	(61.0)[a]	(10.4)	(59.0)	(45.0)	(23.0)
Minerals					
Calcium	3.4 mg/g	49.0	53.0	55.5	54.7
Magnesium	1.8 mg/g	42.8	86.4	67.2	70.2
Potassium	1.6 mg/g	2.2	42.6	47.7	97.8
Iron	84.4 µg/g	84.8	72.7	79.3	78.4
Zinc	36.4 µg/g	77.5	57.4	45.5	33.6
Combined Density Fraction					
Yield	5.0 g	68.0	76.0	70.0	70.0
Protein	54.4 %	76.9	82.6	79.7	83.5
Phytate	55.2 mg/g	39.8	48.1	28.5	44.6
	(71.0)[a]	(10.0)	(55.4)	(68.0)	(34.0)
Minerals					
Calcium	1.0 mg/g	53.6	70.0	62.6	70.0
Magnesium	2.8 mg/g	33.1	77.7	62.9	75.9
Potassium	2.9 mg/g	1.0	51.0	68.5	95.1
Iron	130.4 µg/g	71.3	66.4	73.3	66.0
Zinc	64.5 µg/g	55.9	44.5	34.0	22.5

[a] Values in the parentheses represent percent of phytate present in water-soluble form.

adjusted to pH 5.15 removed 73 and 88% of the phytate, respectively. Dialysis of rapeseed flour dispersion (pH 5.80) without pre-adjustment of pH removed up to 67% of the phytate. Reddy et al.[13] reported that about 40 and 51% of the phytate respectively could be removed from unadjusted dispersions (pH 6.3) of Great Northern bean (GNB) flour and combined density fraction (CDF) during dialysis in a membrane tubing with molecular weight cut-off of 1,000 daltons (Table 6). The maximum amount of phytate was eliminated from the dialyzed dispersions of GNB flour and CDF adjusted to pH 7.0. The greater losses of phytate in GNB flour and CDF was attributed to dissociation of phytate-protein complexes into soluble lower molecular weight phytic acid salts which were removed by dialysis. Okubo et al.[27] established conditions for disassociation of phytate-protein complexes at above and below the isoelectric point of soy globulins for the preparation of low-phytate soy protein isolates. About 95% of phosphorus was removed at pH 5.50 (near isoelectric point of soy globulins) (Table 7). Below isoelectric point (at pH 2.0), only 30% of phosphorus is removed. The remainder may be firmly bound to the protein. It appears that the addition of (EDTA) to soy extracts at high pH values increases phosphorus removal. However, dialysis appears to be impractical on a commercial scale.

Laboratory scale experiments indicate that the ultrafiltration may be promising for selective removal of phytate.[14] Ultrafiltration offers the advantages of mild processing conditions and selectivity. Ultrafiltration is a pressure-activated process that results in a greater rates of removal of solution (i.e., flux).[14] Since phytate and its metal salts have a molecular weight of 1,000 or less,

TABLE 7
Dissociation of Phytate-Protein Complexes in Soybean Extracts at Various pH Values[27]

pH	Extractant and conditions	Phosphorus removed (%)	Phosphorus content[a] (g/100 g protein)
8.5	0.026 M Borate buffer (Na⁺)		
	+ EDTA	80	0.24
	− EDTA	65	0.42
8.5	0.05M Tris-HCl buffer (Cl⁻)		
	+ EDTA	72	0.33
	− EDTA	50	0.60
7.2	0.05M Tris-HCl buffer (Cl⁻)		
	+ EDTA	85	0.18
	− EDTA	82	0.21
5.5	Distilled water		
	+ EDTA	95	0.06
	− EDTA	95	0.06
2.0	0.01N HCl	31	0.82

[a] Original soybean extract used in experiments contained 1.20 g phosphorus per 100 g protein. The extracts were dialyzed for 6 days at 25°C.

use of a membrane with lower molecular weight cut-off may be suitable for producing protein isolates or concentrates with low phytate content. Okubo et al.[27] prepared a low phytate soy protein isolate by first dissociating phytate from protein by chemical treatment and then passing the dissociated phytate through a ultrafiltration membrane (molecular weight cutoff 30,000 daltons), which is permeable to phytate and its metal salts but impermeable to protein. The chemical treatment consisted of either (1) enzymatic hydrolysis by indigenous phytase at pH 5.0, (2) addition of 0.5 M calcium chloride at pH 3.0, or (3) addition of 0.05 M EDTA (Na⁺) at pH 8.5. Little or no removal of phosphorus occurred either at pH 8.5 in the absence of EDTA or at pH 3.0 in the absence of calcium ions (Figure 3). Of the three chemical treatments, the use of EDTA at pH 8.5 and 65°C appears to be least effective in phosphorus removal in comparison to the phytase treatment at pH 5.0 and calcium ion at pH 3.0. All the isolates contained more than 90% protein, but the phosphorus content of the isolate prepared by the EDTA/pH 8.5 treatment was significantly higher than that of other isolates. The two chemical treatments (addition of calcium chloride at pH 3.0, and addition of EDTA [Na⁺] at pH 8.5) require a subsequent washing of retentate with water to remove the calcium or EDTA from the isolate.

Omosaiye and Cheryan[28] studied the feasibility of ultrafiltration (membrane with molecular weight cutoff at 50,000 daltons) to remove phytate from soybean aqueous extracts. They found that the phytate removal from soy concentrate is governed by the environment and state of phytate binding to rejected species. Dilution and reultrafiltration of aqueous soybean extract at pH 6.7 results in about 95% of phytate removal (Table 8). An advantage of the process at pH 6.7 appears to be that no manipulation of pH or chemical treatment is required to produce a protein product with little or no phytate. The removal of phytate from soy products by commercial ultrafiltration would be best achieved by either the addition of cations at low pH or removal of cations at high pH by sequestering agents.[34] These processes may be applicable to other legumes and need further investigations for optimization.

V. ION EXCHANGE PROCESS

Smith and Rackis[31] employed anion-exchange resin alone and a combined application of

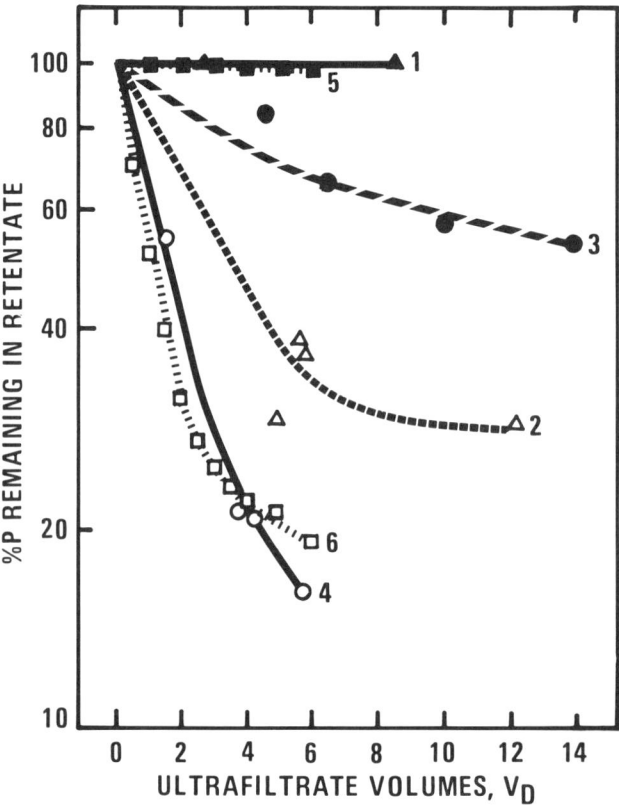

FIGURE 3. Removal of phytate (as measured by phosphorus remaining in retentate) during diafiltration through PM-30 membrane: 1 — pH 8.5, 0.015 M sodium borate, 65°C; 2 — pH 8.5, 0.01 M sodium carbonate/bicarbonate, 65°C, 0.05 M EDTA initially present in the extract; 3 — pH 7.1, sodium carbonate/bicarbonate, 65°C; 4 — pH 5.0, water, 65°C; 5 — pH 3.0, water, 25°C; 6 — pH 3.0, water, 0.5 M calcium chloride, 25°C. (From Okubo, K., Waldrop, A. B., Iacobucci, G. A., and Myers, D., *Cereal Chem.*, 52, 263, 1975. With permission.)

TABLE 8
Composition[a] of Soy Protein Concentrates by Utrafiltration at pH 6.7[28]

Treatment	Protein (%)	Fat (%)	Phytate content (%)	Phytate removed (%)
Original soybeans	43.3	24.0	1.27	—
Water extract	48.3	26.4	1.68	—
Ultrafiltered	56.7	32.5	0.823	35.2
Reultrafiltered	59.7	34.2	0.064	95.0

[a] Expressed on dry weight basis.

dialysis and anion-exchange resin to remove free phytic acid and its salts from water extracts of soybean meal at different pH ranges. They reported that about 78% of the phytate could be removed without appreciable loss of proteins by a combined application of dialysis and anion-

TABLE 9
Removal of Phytate from Soy Protein Extract by Ion-Exchange Process[35]

Isolate	Phytate content (%)	Phytate removed (%)
Control isolate	2.03	—
Cation-exchanged isolate	1.33	34.5
Cation/anion-exchanged isolate	<0.011	>99.0

exchange resin (Dowex-1-X10^R) (Table 5). Recently, Brooks and Morr[35] developed a two-stage ion-exchange process for effective removal of phytate from soy protein extracts. In this process, an alkaline extract of soybeans is passed through cation and anion-exchange resin columns prior to drying. Cation exchange resin treatment removed 97 to 99% of calcium and magnesium and about 34.5% of phytate. They obtained a reduction of 96 to 99% of phytate in soy protein extracts by filtering them through cation and anion-exchange resin columns (Table 9). The maximum amount of phytate was eliminated from soy protein isolates as a result of sequential disruption of the protein-calcium/magnesium-phytate complex by the ion-exchange treatment. Seo and Morr[36] removed more than 85% of phytate from peanut protein extracts during preparation of peanut protein products by ion-exchange treatment. However, commerical applicability of this method has not been realized. An anion-exchange method could be used to reduce the phytate content in soybean oil.[37]

VI. ENZYME TREATMENT

Several researchers[38-41,45] have studied the hydrolysis of phytate by phytase of plant and microbial sources. This section has been discussed extensively in the previous chapter. Chang et al.[42] investigated the hydrolysis of phytate in California small white beans by phytase. They used several incubation conditions that included (1) incubation of beans in buffer (pH 5.5) at 50°C for 24 h, (2) incubation of pre-soaked beans in water at 60°C for 10 h, and (3) exposure of whole beans to water-saturated air at 60°C. About 33 to 90% of phytate was eliminated from California small white beans by these conditions, and these conditions caused similar removal of phytate from mung beans, lima beans, and wheat. Anno et al.[43] eliminated most of the phytate in commercial soybean milk by treating it with wheat phytase. The enzyme treatment appears to be effective in reducing phytate in cereals, legumes, and their processed food products.

VII. GENETIC SELECTION

The genetic approach for the selection of low-phytate wheat varieties has been suggested[44] and may be extended to other cereals and legumes. However, at the present time, genetic approach may not be successful due to lengthy procedures involved in it.

There are no practical methods to remove phytate completely from cereals, legumes, and their products without altering either the bioavailability of nutrients or the functional and sensory characteristics.[5,14] Isolation and characterization of the precise chemical form(s) in which phytic acid occurs in cereals and legumes may provide additional information for development of commercial removal methods.

REFERENCES

1. **de Rham, O. and Jost, T.,** Phytate-protein interactions in soybean extracts and low-phytate soy protein products, *J. Food Sci.,* 44, 596, 1979.
2. **Hartman, G. H.,** Removal of phytate from soy protein, *J. Am. Oil Chem. Soc.,* 56, 731, 1979.
3. **Reddy, N. R., Sathe, S. K., and Salunkhe, D. K.,** Phytates in legumes and cereals, *Adv. Food Res.,* 28, 1, 1982.
4. **Morris, E. R.,** Phytate and dietary mineral bioavailability, in *Phytic acid: Chemistry and Applications.,* Graf, E., Ed., Pilatus Press, Minneapolis, 1986, 57.
5. **Harland, B. F. and Oberleas, D.,** Phytate in foods, *World Rev. Nutr. Diet.,* 52, 235, 1987.
6. **Thompson, L. U., Button, C. L., and Jenkins, D. J. A.,** Phytic acid and calcium affect the *in vitro* rate of navy bean starch digestion, and blood glucose response in humans, *Am. J. Clin. Nutr.,* 46, 467, 1987.
7. **O'Dell, B. L., deBoland, A. R., and Koirtyohann, S. R.,** Distribution of phytate and nutritionally important elements among the morphological components of cereal grains, *J. Agric. Food Chem.,* 20, 718, 1972.
8. **Deshpande, S. S., Sathe, S. K., Salunkhe, D. K., and Cornforth, D. P.,** Effects of dehulling on phytic acid, polyphenols, and enzyme inhibitors of dry beans (*Phaseolus vulgaris* L.), *J. Food Sci.,* 47, 1846, 1982.
9. **Honig, D. H., Wolf, W. J., and Rackis, J. J.,** Phytic acid and phosphorus content of various soybean protein fractions, *Cereal Chem.,* 61, 523, 1984.
10. **Honig, D. H., Rackis, J. J., and Wolf, W. J.,** Effect of pH and salt on yields, trypsin inhibitor content, and mineral levels of soybean protein isolates and wheys, *J. Agric. Food Chem.,* 35, 967, 1987.
11. **Naczk, M., Rubin, L. J., and Shahidi, F.,** Functional properties and phytate content of pea protein preparations, *J. Food Sci.,* 51, 1245, 1986.
12. **Honig, D. H. and Wolf, W. J.,** Mineral and phytate content and solubility of soy protein isolates, *J. Agric. Food Chem.,* 35, 583, 1987.
13. **Reddy, N. R., Sathe, S. K., and Pierson, M. D.,** Removal of phytate from Great Northern beans (*Phaseolus vulgaris* L.) and its combined density fraction, *J. Food Sci.,* 53, 107, 1988.
14. **Cheryan, M.,** Phytic acid interactions in food systems, *CRC Crit. Rev. Food Sci. Nutr.,* 13, 297, 1980.
15. **Reddy, N. R.,** Milling and Biochemical Characteristics of Triticale, M. S. thesis, Alabama A & M University, Normal, Alabama, 1976.
16. **Tabekhia, M. M. and Donnelly, B. J.,** Phytic acid in durum wheat and its milled products, *Cereal Chem.,* 59, 105, 1982.
17. **Ressurreccion, A. P., Juliano, B. O., and Tanaka, Y.,** Nutrient content and distribution in milling fractions of rice grain, *J. Sci. Food Agric.,* 30, 475, 1979.
18. **Nayini, N. R. and Markakis, P.,** Effect of milling extraction on the inositol phosphates of wheat flour and bread, *J. Food Sci.,* 48, 1384, 1983.
19. **Tabekhia, M. M. and Luh, B. S.,** Effect of Milling on macro and micro-minerals and phytate of rice, *Dtsch. Lebensm.-Rundsch.,* 75, 57, 1979.
20. **Elkowicz, K. and Sosulski, F. W.,** Antinutritive factors in eleven legumes and their air-classified protein and starch fractions, *J. Food Sci.,* 47, 1301, 1982.
21. **Tecklenburg, E., Zabik, M. E., Uebersax, M. A., Dietz, J. C., and Lusas, E. W.,** Mineral and phytic acid partitioning among air-classified bean flour fractions, *J. Food Sci.,* 49, 569, 1984.
22. **McKinney, L. L., Sollars, W. F., and Setzkorn, E. A.,** Studies on the preparation of soybean protein free from phosphorus, *J. Biol. Chem.,* 178, 117, 1949.
23. **Goodnight, K. C., Hartman, G. H., and Marquardt, R. F.,** Aqueous purified soy protein and beverage, U.S. Patent, 3,995,071, 1976.
24. **Ford, J. R., Mustakas, G. C., and Schmutz, R. D.,** Phytic acid removal from soybeans by a lipid protein concentrate process, *J. Am. Oil Chem. Soc.,* 55, 371, 1978.
25. **Baker, E. C., Mustakas, G. C., Erdman, J. W., Jr., and Black, L. T.,** The preparation of soy products with different levels of native phytate of zinc bioavailability studies, *J. Am Oil Chem. Soc.,* 58, 541, 1981.
26. **Hill, R. and Tyler, C.,** The reaction between phytate and protein, *J. Agric. Sci.,* 44, 324, 1954.
27. **Okubo, K., Waldrop, A. B., Iacobucci, G. A., and Myers, D. V.,** Preparation of low-phytate soy protein isolate and concentrate by ultrafiltration, *Cereal Chem.,* 52, 263, 1975.
28. **Omosaiye, O. and Cheryan, M.,** Low-phytate, full fat soy protein product by ultrafiltration of aqueous extracts of whole soybeans, *Cereal Chem.,* 56, 58, 1979.
29. **Brooks, J. R. and Morr, C. V.,** Phytate removal from soybean proteins, *J. Am. Oil Chem. Soc.,* 61, 1056, 1984.
30. **Deshpande, S. S. and Cheryan, M.,** Preparation and antinutritional characteristics of dry bean *(Phaseolus vulgaris* L.), protein concentrates, *Qual. Plant. Plant Foods Human. Nutr.,* 34, 185, 1984.
31. **Smith, A. K. and Rackis, J. J.,** Phytin elimination in soybean protein isolation, *J. Am. Chem. Soc.,* 79, 633, 1957.
32. **McKinney, L. L. and Sollars, W. F.,** Extraction of soybean protein with sulfurous acid, *Ind. Eng. Chem. (Anal. Ed.),* 41, 1058, 1949.

33. **Serraino, M. R. and Thompson, L. U.**, Removal of phytic acid and protein-phytic acid interactions in rapeseed, *J. Agric. Food Chem.*, 32, 38, 1984.
34. **Prattley, C. A., Stanley, D. W., and Van De Voort, F. R.**, Protein-phytate interactions in soybeans, II. Mechanism of protein-phytate binding as affected by calcium, *J. Food Biochem.*, 6, 255, 1982.
35. **Brooks, J. R. and Morr, C. V.**, Phytate removal from soy protein isolates using ion-exchange processing treatments, *J. Food Sci.*, 47, 1280, 1982.
36. **Seo, A. and Morr, C. V.**, Activated carbon and ion-exchange treatments for removing phenolics and phytate from peanut protein products, *J. Food Sci.*, 50, 262, 1985.
37. **Winters, D. D., Handel, A. P., and Lohrberg, J. D.**, Phytic acid content of crude, degummed, and retail soybean oils and its effect on stability, *J. Food Sci.*, 49, 1113, 1984.
38. **Kon, S., Olson, A. C., Frederick, D. P., Eggling, S. B., and Wagner, J. R.**, Effect of different treatments on phytate and soluble sugars in California Small White beans, *J. Food Sci.*, 39, 215, 1973.
39. **Becker, R., Olson, A. C., Frederick, D. P., Kon, S., Gumbamann, M. R., and Wagner, J. R.**, Conditions for autolysis of alpha-galactosides and phytic acid in California Small White beans, *J. Food Sci.*, 39, 766, 1974.
40. **Ranhotra, G. S., Loewe, R. J., and Puyat, L. V.**, Phytic acid in soy and its hydrolysis during breadmaking, *J. Food Sci.*, 39, 1023, 1974.
41. **Reinhold, J. G.**, Phytate destruction by yeast fermentation in whole wheat meals, *J. Am. Diet. Assoc.*, 66, 38, 1975.
42. **Chang, R., Schwimmer, S., and Burr, H. K.**, Phytate: removal from whole dry beans by enzymatic hydrolysis and diffusion, *J. Food Sci.*, 42, 1098, 1977.
43. **Anno, T., Nakanishi, J., Matsudno, R., and Kamikubo, K.**, Enzymatic elimination of phytate in soy milk, *Nippon Shokuhin Kogyo Gakkai-Shi*, 32, 174, 1985.
44. **Bassiri, A. and Nahapetian, A.**, Influences of irrigation regiments on phytate and mineral contents of wheat grain and estimates of genetic parameters, *J. Agric. Food Chem.*, 27, 984, 1979.
45. **Han, Y. W. and Wilfred, A. G.**, Phytate hydrolysis in soybean and cotton seed meals by *Aspergillus ficuum* phytase, *J. Agric. Food Chem.*, 36, 259, 1988.
46. **Thompson, L. U.**, Reduction of phytic acid concentration in protein isolates by acylation techniques, *J. Am. Oil Chem. Soc.*, 64, 1712, 1987.

INDEX

A

Accumulation of phytates, 7, 41
Acid phosphomonoesterase, 19
Activation energies, 16—17
Adenosine diphosphate (ADP)-phosphotransferase, 9
ADP, see Adenosine diphosphate
Aflatoxins, 23, see also specific types
Albumin, 62, 99, see also specific types
Aleurone particles (globoids), 11, 16, 23, 39
Alkaline phosphatase, 72
Amperometry, 28, 35
α-Amylase, 100
Analysis of phytate, 27—35, see also specific methods
  qualitative methods of, 27—28
  quantitative methods for, 28—29
    chromatographic methods, 31—35
    nuclear magnetic resonance, other methods and, 35
    precipitation methods, 29—31
Animals, 81, see also specific types
  calcium availability in, 96
  digestion of phytates in, 71—75
  iron availability in, 92—94
  magnesium availability in, 98—99
  zinc availability in, 82—86
Anion-exchange resin, 144
Antibiotics, 25, see also specific types
Antioxidant property of phytates, 23, see also specific types
Arginine, 57
Ascorbic acid, 24
Assays, 82, see also specific types
Atomic absorption, 66
Autolysis, 126—132
Autoxidation, 24
Availability
  of calcium, 96—98
  of iron, 92—96
  of magnesium, 98—99
  of minerals, 81
  of phytates, 71—76
  of zinc, see under Zinc

B

Bioassays, 82, see also specific types
Bioavailability, see Availability
Biosynthesis of phytates, 11, 15—19
Bone resorption, 25
Bone turnover, 25
Breadmaking, 120—126

C

Cadmium, 68
Calcium, 17—18, 57, 59, 60, 66, 67
  absorption of, 81
  availability of, 96—98
Calcium chloride, 111
Calcium magnesium, 23
Calcium phytate, 3, 25, 68
Carbohydrates, 40, see also specific types
Cariogenesis, 25
Cariostatic properties, 25, see also specific types
Casein, 81, 100
Cations, 23, see also specific types
Cattle, 71
CDF, see Combined density fraction
Chemical structure of phytates, 1
Chicks, 81
p-Chloromercuribenzoate, 18
Chromatography, 31—35, see also specific types
  high-performance liquid (HPLC), 29, 33—35
  ion exchange, 28, 31—33
  paper, 27
  Sephadex CM-50, 16
  thin-layer (TLC), 28
Cobalt, 17—18
Colonic carcinogenesis, 23
Colon tumors, 24
Combined density fraction (CDF), 66, 142
Commercial manufacture of phytates, 24—25
Complexometric titration, 28
Content of phytates, 41—47, see also specific types
Cookability, 23, 111
Cooking, 111—115
Copper, 66, 68, 99
Coprecipitation of phytates, 61
Coronary heart disease, 24
Cycloheximide, 11—12

D

Degradation of phytates, 12
Dephosphorylation, 7, 16—17
Dietary intake of phytates, 47—50
Differential solubility methods, 137—141, see also specific types
Digestion
  of phytates, 71—76, 81
  of protein, 100—102
  of starches, 100—102
Direct precipitation, 29, 31
Dissociation constants for phytates, 5
Distribution of phytates, 41, 42
Divalent metal ions, 17, 57, see also specific types
Dogs, 94
Dormancy, 23

E

EDTA, see Ethylenediaminetetraacetate
Egg albumin, 99
Electron-dense aleurone particles, 39

Electrophoresis, 16, 27—28, see also specific types
Enamel dissolution, 25
Energies, see also specific types
   activation, 16—17
   store of, 23
Environmental factors, 47, see also specific types
Enzymes, 25, see also specific types
   inhibition of, 100
   kinase, 8
   removal of phytates and, 145
Ethylenediaminetetraacetate (EDTA), 27, 32, 35, 142, 143
N-Ethylmaleimide, 18

F

Fermentation, 25, 120—126
Ferric phytate, 30
Fiber, 81
Fluoride, 17
Food applications of phytates, 24
Formation of phytates, 7
Fourier transform-nuclear magnetic resonance (FT-NMR), 29, 35
FT-NMR, see Fourier transform-nuclear magnetic resonance

G

β-Galactosidase, 100
Gel filtration, 59
Genetic selection for removal of phytates, 145
Germination, 115—120
Gibberellic acid, 12
Globoids, see Aleurone particles
Glucose-6-phosphate, 7, 11

H

HCl, see Hydrochloric acid
High-performance liquid chromatography (HPLC), 29, 33—35
High pH phytate-protein interactions, 59
Histidine, 57
Hormones, 12, see also specific types
HPLC, see High-performance liquid chromatography
HSA, see Human serum albumin
Humans
   calcium availability in, 96—98
   iron availability in, 95—96
   magnesium availability in, 99
   zinc availability in, 86—88
Human serum albumin (HSA), 62
Hydrochloric acid (HCl), 31, 32
Hydrolysis, 10—12, 16, 19, 24, 81
   increase in, 121
   inhibition of, 121, 128
   maximization of, 127
   in seeds, 115
   yeast and, 121

I

Indirect precipitation, 29—31
Inflammatory bowel diseases, 23
Inorganic phosphates, 11, 115, see also specific types
Inorganic phosphorus, 126
Inositol, 28, 115
Intermediate pH phytate-protein interactions, 58—59
Iodoacetamide, 17
Ion exchange chromatography, 28, 31—33
Ion exchange process, 143—145, see also specific types
Iron, 17—18, 81, 92—96
Iron chelating properties of phytates, 25
Iron phytate, 30
Isotachophoresis, 35

K

Kidney stones, 24, 25
Kinases, 8, see also specific types

L

Lactic acid, 25
Lead, 68
Lead poisoning, 25
Lipid peroxidation, 23
Low pH phytate-protein interactions, 57

M

Magnesium, 17—18, 40, 81, 98—99
Magnesium chloride, 111
Manganese, 99
Manufacture of phytates, 24—25
Mechanical processes, 137, see also specific types
Medical applications of phytates, 24, 25
Membrane separation for removal of phytates, 141—143
Metabolic cycle of phytates, 9
Metabolism of phytates, 10
Metals, 100, see also specific types
   absorption of, 25
   binding of, 68
   divalent, 17, 57
Metal salts, 41, see also specific types
Minerals, 50, see also specific types
   bioavailability of, 81
   phytate effects on, 81—100
      calcium, 96—98
      iron, 92—96
      magnesium, 98—99
      other minerals, 99—100
      zinc, 81—82
         animal studies, 82—86
         human studies, 86—88
         phytate/zinc molar ratio for predicting zinc bioavailability, 88—92
   phytate interactions with, 66—68

protein interactions with, 102
Myoinositol, 23
Myoinositol-1,3-diphosphate, 9, 10
Myoinositol-2,4-diphosphate, 10
Myoinositol-1,2,3,4,5,6 hexakisphosphate, 15
Myoinositol-1,2-monophosphate, 8
Myoinositol oxidation pathway, 23
Myoinositol-1,2,3,4,5 pentakisphosphate, 15
Myoinositol pentaphosphate, 9
Myoinositol-1,2,3,4,5-pentaphosphate, 10
Myoinositol-1,2,4,5,6-pentaphosphate, 10
Myoinositol-1,3,4,5,6-pentaphosphate, 9—11
Myoinositol-1-phosphate, 8, 9, 11
Myoinositol-2-phosphate, 7, 9—11
Myoinositol-1-phosphate synthase, 7
Myoinositol-1,2,3,4,5-tetraphosphate, 9
Myoinositol-1,2,4,5-tetraphosphate, 10
Myoinositol-1,3,5-triphosphate, 9
Myoinositol-2,4,5-triphosphate, 10

N

NaCl, see Sodium chloride
NMR, see Nuclear magnetic resonance
Nuclear magnetic resonance (NMR), 4, 9, 29, 35
Nutritional consequences of phytates, 81—102
  enzyme inhibition and, 100
  minerals and
    calcium, 96—98
    iron, 92—96
    magnesium, 98—99
    other minerals, 99—100
    zinc, 81—82
      animal studies, 82—86
      human studies, 86—88
      phytate/zinc molar ratio for predicting zinc bioavailability, 88—92
  protein digestibility and, 100—102
  starch digestibility and, 100—102

O

Occurrence of phytates, 39—41
Orthophosphate, 15, 115
Ovalbumin, 62
Oxalates, 81, see also specific types

P

Paper chromatography, 27
Paper electrophoresis, 27—28
Pepsin, 100
PER, see Protein efficiency ratio
Peroxidation, 24
Phosphates, 11, 115, see also specific types
Phosphorus, 23, 59, 61
  inorganic, 126
  store of, 23
Phosphorylation, 7—10, see also specific types
Physiological functions of phytates, 23—24

Phytases, 28, 121, see also specific types
  activation of, 12
  activators of, 17—18
  biosynthesis of, 15—19
  characterization of, 15—19
  extraction of, 16
  inhibitions of, 11
  location of, 16
  molecular properties of, 16—19
  purification of, 16
  in seeds, 16
  specificities of, 18
  synthesis of, 16
Phytate-protein interactions, 57—66, 102
  complexes, formation of,
    high pH, 59
    intermediate pH, 58—59
    low pH, 57
  with other proteins, 61—62
  protein functionality, effects on, 62—66
  with soy proteins, 59—61
Phytate/zinc molar ratio for predicting zinc availability, 88—92
Plaque formation, 25
Polyphenolic compounds, 81
Polysaccharides, 23, 81, see also specific types
Potassium, 40
Potassium phytate, 41, 111
Poultry, 72—74
Precipitation methods, 29—31, see also specific types
Precursors of phytates, 7, see also specific types
Processing technologies, see specific types
Properties of phytates, 1, see also specific types
Protein efficiency ratio (PER), 64
Proteins, 40, 41, see also specific types
  digestibility of, 100—102
  functionality of, 62—66
  mineral interactions with, 102
  soy, 59—61, 102
  unbound, 102
Putrefaction, 23

Q

Qualitative methods for analysis of phytate, 27—28, see also specific types
Quantitative methods for analysis of phytate, 28—29
  chromatographic methods, 31—35
  nuclear magnetic resonance, other methods and, 35
  precipitation methods, 29—31

R

Rabbits, 71
Raman spectroscopy, 4
Rats, 74—75, 81, 82, 94, 99
Removal of phytates, 137—145, see also specific methods
  differential solubility methods for, 137—141

enzyme treatment for, 145
genetic selection for, 145
ion exchange process for, 143—145
mechanical processes for, 137
membrane separation for, 141—143
selective extraction for, 137—141
Renal calculi, 24, 25
Resins, 144, see also specific types

S

Selective extraction for removal of phytates, 137—141
Selenium, 99, 100
Sephadex CM-50 chromatography, 16
Sheep, 71
Soaking, 126—132
Sodium chloride (NaCl), 31, 33
Sodium cyclotetrametaphosphate tetrahydrate, 35
Sorbic acid, 24
Soy proteins, 59—61, 102, see also specific types
Spectroscopy, see also specific types
    Fourier transform-NMR, 29, 35
    nuclear magnetic resonance (NMR), 4, 9, 35
    Raman, 4
Starches, 100—102, see also specific types
Stepwise phosphorylation, 7, 8, 10
Structure of phytates, 1, 3, see also specific types
Swine, 72, 81
Synthesis of phytates, 11, 15—19

T

TCA, see Trichloroacetic acid
Thin-layer chromatography (TLC), 28
Titration, 35, see also specific types
    complexometric, 28
    turbidimetric, 59
TLC, see Thin-layer chromatography
Trichloroacetic acid (TCA), 32, 35
Trypsin, 100, 102
Tumors, 24, see also specific types

U

Ultrafiltration, 142
Unbound proteins, 102, see also specific types
Uses of phytates, 24—25, see also specific types

V

Vitamin C, 24
Vitamin D, 72

X

X-ray crystallography, 4

Y

Yeast, 121

Z

Zinc, 66, 68, 81
    availability of, 81—92
        in animals, 82—86
        in humans, 86—88
        phytate/zinc molar ratio for predicting, 88—92
        seed maturity and, 84
    deficiency in, 81, 88